The Unlimited Power of Your Mind and Body

How to Live Longer Naturally by Reprogramming Your Mind, Body, and Genes for Strength and Vitality

HEALTH AND LONGEVITY MASTERY SERIES
VOLUME FIVE

By Tad Sisler

This book is dedicated to my Uncle Ted Witt, who mentored me and opened my eyes to life's possibilities beyond what we see… and to my sister Judy, who understood consciousness on levels I'm still trying to understand… I love you both!

TABLE OF CONTENTS

FOREWORD

This is the fifth book in my **Health and Longevity Mastery Series**, and I'm very excited to present it to you. This book includes ideas, concepts, and scientific evidence from each of my other books in this series and builds on the premise that we are on a new frontier of discovery unlike any other in human history. In this book, I will address physical, mental, emotional, spiritual, and technological aspects of longevity. In essence, I focus on natural methods for living longer while exploring the magic of human potential.

Some inventions since the beginning of time overshadow others: the invention of the wheel, the Printing Press, the harnessing and distribution of electricity, vaccinations, the internet.

Even since I was born, we've placed men on the moon and spaceships on other planets; we've witnessed the invention of the smartphone where we hold all the knowledge of the universe in our pocket; we sequenced the human genome; CRISPR and Genetic Engineering opened up doors to potential cures for diseases, revolutionary cancer treatments, and the engineering of more resilient crops; and now, AI and machine learning are about to explode us all into a new Renaissance, where we will witness four hundred years of progress within the next ten years. Who knows what new "inventions" will be revealed to us (because, really, aren't inventions just rediscovered ideas that were always there but we hadn't figured them out yet?), and how will that relate to the as of yet undiscovered power within our minds and bodies?

This book will answer some of those questions and raise new ones. This is good, because two of the most potent powers we have are the ability to provoke thought, and the power of choice. We can add years to our lives by choosing to be optimistic, as I evidenced in my book **The Science of Positive Thinking: How Mindset, Daily Habits, and Emotional Well-being Can Add Years to Your Life.** We can choose to embrace new information and methodology, and so I'll be unafraid to present some revolutionary ideas. After all, when my grandmother **Audrey Athey Sisler** was born in 1889, the entire world would have thought it was preposterous for humans to actually build a flying machine. In her 94 years, she went from horse and buggy to spaceships; from the telegraph to the personal computer; from debilitating diseases to a clean water supply and penicillin; and the list goes on. Just wait and behold what's about to unfold with the advent of AI and quantum computers (and quantum computers are engineered using quantum mechanics).

"If quantum mechanics hasn't profoundly shocked you, you haven't understood it yet." – Neils Bohr

Throughout this book, I will also unfold action plans within each chapter and in the Appendices at the end of the book. I will give you a road map to maximize the potential of your incredible **mind** and **body**, and I know you will do great things with it. If you've read the first four books of my **Health and Longevity Mastery Series** in order, this book will pull it all together for you.

If you haven't done this, this book stands alone and you can always go back later to my other books where you'll learn about cutting-edge longevity science for reversing aging... or how to boost your immunity and increase energy by using vitamins, supplements and herbs... or how you can achieve fast, lasting weight management with a personalized, modern approach... or how your mindset, daily habits, and emotional well-being can thrive through a positive outlook.

Wait until you see what the next ten years unfolds! I love the feeling that we're on the edge of a new Golden Age, a new universe of possibilities, and I can't wait to give you a peek into it.
It all starts in our minds. Remember the words of **Napoleon Hill:**

"What the mind can conceive and believe, it can achieve."

Napoleon Hill
Credit – FLICKR / creativecommons.org

Yes, we can achieve whatever we can conceive and believe. But we must have an open mind to get there. Most of the world in 1492 believed strongly that Christopher Columbus would sail off the edge of the earth and never return. Edison was laughed at as he failed through hundreds of experiments before inventing the light bulb. They kept an open mind, stayed firm in their beliefs, and expanded our knowledge and awareness by doing so.

We are all products of our unique history. Childhood experiences, belief systems, mentors, even prejudices have influenced our outlook and behavior, sometimes without us even knowing it. Sometimes it's hard, but we must keep an open mind as we explore new possibilities.

We will now jump into the world of the greatest invention of all, the human brain and body. And remember, quantum computing will quickly be far more intelligent than the human brain, but it will never quite know the power of our soul, our empathy, our emotional power. This is what separates us from the machines. Imagine humans as half-animal, half-god (I'll get more into this concept in Chapter 13). Our instincts work with our intellect to make us extraordinarily unique within species. And we haven't even discovered half of our real potential yet. Let's find out more!

"I think self-discovery is the greatest achievement in life, because once you discover yourself and accept what you are, then you can fulfil your true potential and be happy." – Marco Pierre White

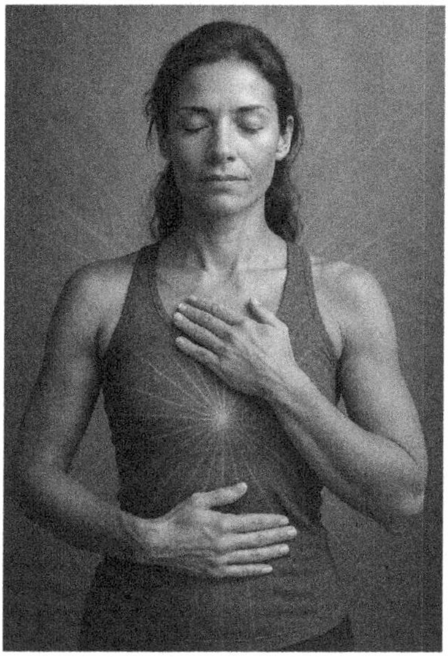

INTRODUCTION
UNLOCKING YOUR INNER POTENTIAL

You've got an incredible power inside you, waiting to be unlocked. This book introduces the "Forever Switch"—a way to tap into your body's natural ability to boost health and live longer. It's not just about rare moments of strength; it's about making vitality a part of your everyday life.

Angela, a mom from Arizona, found herself facing a parent's worst fear. When her son was trapped under a two-ton SUV, this 128-pound woman lifted the vehicle just enough to save him: no tricks, no tools—just raw, extraordinary strength (*BBC Future*).

You might think **Angela's** feat was just adrenaline, but research from places like Harvard and Stanford shows we all have hidden reserves of strength, healing, and even longevity. These studies suggest that we can access these abilities far beyond what we normally expect with the right approach.

Take **Uri Geller**, who famously bent spoons with his mind during experiments at Stanford in the 1970s. Under strict conditions, he showed that mental focus could achieve things that seem impossible (*Mind-Reach*).

Wim Hof, aka "The Iceman," is a Dutch guy who can handle crazy cold like it's no big deal—think chilling in an ice bath for nearly 2 hours in 2007, setting a world record at 1 hour, 52 minutes, and 42 seconds, without even getting hypothermia! He's also climbed most of Mount Everest in just shorts and boots, hitting 7,400 meters before a foot injury stopped him, not the freezing temps. His secret? The *Wim Hof Method*—breathing tricks, cold exposure, and mental focus—that lets him control his body in ways science didn't think possible, like lowering inflammation and boosting adrenaline to superhuman levels, showing we might all have a hidden toughness to unlock. I reference the *Wim Hof Method Study* in Chapter 12.

Dean Karnazes, an American ultramarathon runner, took endurance to a whole new level by running 350 miles across California in 2005 without stopping—80 hours straight, no sleep, just eating on the go and pushing through pure grit for over three days! He also ran 50 marathons in 50 states in 50 days back in 2006, then ran back to the start, covering 1,310 miles total. Researchers found his body clears lactate—a fatigue-causing byproduct—way better than most, letting him keep going when others would crash, proving how far mental toughness and physical adaptation can take us with the right training (Lactate Clearance in Endurance).

Malala Yousafzai, a Pakistani activist, showed insane resilience after the Taliban shot her in the head at 15 in 2012 for fighting for girls' education—she survived, woke up from a coma, and kept right on advocating! By 2013, she was speaking at the UN, later wrote a bestselling book, *I Am Malala,* and became the youngest Nobel Peace Prize winner in 2014 at 17, all while facing death threats. Her story proves our "god-self"—that mental strength to find purpose—can turn even the worst trauma into a force for good, inspiring us all to dig deep and keep going no matter what (*Malala's Recovery and Advocacy*).

"Let us make our future now, and let us make our dreams tomorrow's reality." – Malala Yousafzai

Malala Yousafzai
Credit – Wikimedia Commons

Then there's **Edgar Cayce**, the "Sleeping Prophet," whose 14,000 trance readings offered eerily accurate diagnoses and predictions, sparking ongoing research into human potential (*There Is a River*). I could give you many more examples of ordinary humans doing super-human feats. Do we all have extraordinary powers within us? How deep does our inner strength go?

This book is your roadmap to activating your "Forever Switch" every day, not just in emergencies. Through practical, science-backed methods, you'll learn how to harness your inner strength for a life filled with energy, clarity, and lasting health.

SECTION ONE: WHY THIS BOOK EXISTS

INTRO.1.1: HIDDEN RESERVES ARE REAL

Did you know your body has built-in limits to how strong you can be? It's true! Your nervous system acts like a governor, keeping your muscle strength to about 40% of its full potential to prevent injuries like torn muscles or broken bones. This idea comes from the central governor theory, which says your brain regulates effort to keep you safe (*Wikipedia: Central governor*).

In emergencies, your adrenaline can override these limits, letting you tap into more power, while mitochondria, the tiny energy factories in your cells, provide the fuel for your muscles to work. This shows you've got incredible strength waiting to be used, but your body saves it for when you really need it. This override probably goes back to the days of cavemen when a huge beast came out of nowhere and an extra burst of strength was needed immediately to survive.

When you're in a dangerous situation, your body can decide it's time to go all out. Those nervous system governors loosen up, allowing you to access strength you didn't know you had. This is why people can do things like lift cars to save someone trapped underneath—it's proof that this amazing capacity is already built into you, just waiting for the right moment to kick in.

Researchers at top places like Harvard and Stanford are discovering ways to unlock more than just physical strength. They're finding methods to boost your immune system, speed up healing, and even slow down aging. For example, a study from Harvard and MIT found a chemical way to make cells act younger, which could help you age more slowly (*SciTechDaily: Age Reversal Breakthrough*). Plus, other things like meditation can strengthen your immune system by reducing stress (*Harvard Health: Boost Your Immune System*). It's like finding secret codes to make your body perform at its best.

Your body is like a super advanced computer where everything's connected. Your brain talks to your gut, affecting your mood and health through what's called the brain-gut axis (*Harvard Health: Gut-Brain Connection*).

Hormones control your energy, stress, and more, while your genes can be turned on or off by your lifestyle through epigenetics (*NIH: Epigenetics and Lifestyle*). Getting how these parts work together is important for staying healthy and living longer. This book is your user manual for all this. It'll walk you through how to make the most of your body's systems—brain, gut, hormones, and genes—so you can feel stronger, healthier, and ready to live a longer, more vibrant life.

INTRO.1.2: A WHOLE-SYSTEM BLUEPRINT

In this book, we take a holistic approach to your health, seeing your body and mind as parts of one big ecosystem. That means your physical strength (muscles), your mental state (mindset), the bacteria in your gut (microbes), and your relationships (family ties) all interact and influence each other. For instance, regular exercise not only builds muscle but can also improve your mood (*Harvard Health*) and support a healthy gut microbiome (*Gut Microbes*). Similarly, a positive mindset can enhance physical health, and strong social connections can reduce stress and promote longevity (*APA*). By understanding these connections, you can make informed choices to optimize your overall well-being.

We're combining the best of both worlds: timeless practices from ancient cultures and the latest advancements in technology. Take ikigai, for example—a Japanese concept that means finding your purpose in life. Studies show that having a strong sense of purpose can lead to a longer, healthier life (*JAMA Psychiatry*).

Then there's stoic calm, inspired by Stoic philosophy, which teaches emotional resilience and has been linked to better mental health (I touch on both concepts in my book **The Science of Positive Thinking: How Mindset, Daily Habits, and Emotional Well-being Can Add Years to Your Life**).

"You have power over your mind—not outside events. Realize this, and you will find strength." – Marcus Aurelius

Marcus Aurelius
Credit – Chat GPT Image Generated

Forest bathing, or shinrin-yoku, involves spending time in nature and has been proven to reduce stress and boost immunity (*International Journal of Environmental Research and Public Health*). On the tech side, AI health assistants can provide personalized health recommendations, and epigenetic clocks can measure your biological age, helping you track your progress (*Nature*). By integrating these, you get a comprehensive approach to enhancing your health and longevity.

To make sure you can apply what you learn, each chapter concludes with two practical tools. The **72-Hour Jump-Start** is a quick action plan that lets you start experiencing benefits right away, within just three days. It's designed to give you a taste of the changes you can make. Then, the **Lifelong Integration** roadmap provides a step-by-step guide to incorporating these practices into your daily routine for the long term. This way, you're not just reading about health improvements; you're actively making them happen. It would mean the world to me to be able to help you live a longer, healthier life with these stepping stones.

We understand that scientific research can be dense and full of technical terms. That's why we've taken the latest peer-reviewed studies and translated them into everyday language. Think of it as having a conversation with a friend over coffee, where complex ideas are explained simply and clearly. You'll get all the important information without getting bogged down by too much scientific jargon, making it mostly easy to understand and apply to your life.

My ultimate aim is to provide you with routines that are realistic and fit seamlessly into your busy life. I'm not offering impractical theories or wishful thinking; instead, you'll find evidence-based strategies that you can implement today, no matter how packed your schedule is. These are down-to-earth solutions designed to help you live longer and healthier without requiring drastic lifestyle changes.

INTRO.1.3: EVIDENCE FIRST, HYPE-FREE

This book is built on solid ground, pulling from top-tier scientific journals like the *New England Journal of Medicine*, which sets the bar for medical research, *Aging Cell*, which digs into the nuts and bolts of aging, and *Rejuvenation Research*, which looks at ways to stay youthful longer. I skip the shaky claims you'd find on influencer blogs or social media, giving you advice you can trust because it's backed by real, peer-reviewed studies.

Think of these case studies as stories from people who've been in your shoes, trying out health strategies to see what works.

They share the wins—like better energy or improved mood—but also the bumps along the way, like habits that didn't stick. This balanced view helps you cut through the hype of miracle cures and focus on what's actually effective for real people.

Want to see results for yourself? I've included quick-win checklists that let you track your own health stats, like how well you're sleeping, your heart rate variability (HRV)—a measure of how your body handles stress—or your blood sugar levels. These simple guides show you how to use apps or devices to measure changes, so you can prove the strategies are working for you.

Health science is always moving forward, and this book is designed to keep up. I've left room to add new stories and breakthroughs as they come, making it a living resource. Whether it's a new study or a fresh example, these placeholders ensure you're always getting the latest insights to support your journey to a longer, healthier life.

My commitment is that you'll finish this book feeling ready to take on your health, not buried under a pile of confusing advice. Everything is laid out in a clear, doable way, with practical steps you can start right away. You'll walk away with the tools and confidence to make lasting changes without feeling stressed or lost.

"There are risks and costs to action. But they are far less than the long-range risks of comfortable inaction." – John F. Kennedy

SECTION TWO: HOW TO USE THIS BOOK

INTRO.2.1: THE 72-HOUR JUMP START

You can use this book purely for information, and if you do, you'll learn amazing things about how our mind and body work! Or you can use it as it is meant for you... a full-on action plan to maximize the efficiency of your mind and body.

In this book, every chapter gives you a simple action that you can start doing right away—within just three days—and you won't need a bunch of fancy equipment to get going. Whether it's a quick exercise routine, a mental health practice, or a dietary change, these actions are designed to be easy to implement. For example, you might be asked to try a 10-minute meditation session each morning or to swap out one unhealthy snack for a healthier option. The idea is to make it so straightforward that anyone can jump in and start seeing benefits almost immediately.

These quick actions aren't just about getting started; they're also about building momentum. When you achieve these small wins, your brain releases dopamine, that feel-good chemical that makes you want to keep going. It's like getting a little reward each time you complete a task, which motivates you to tackle the more in-depth material later on.

Research shows that breaking down big goals into smaller, achievable steps can lead to greater success and sustained effort (*American Psychologist*). So, by starting with these quick wins, you're setting yourself up for long-term success.

The main thing is, no matter what you do, work it with gusto! My dear friend, legendary *NFL* Hall-of-Fame linebacker **Junior Seau** had one of the strongest work ethics I've ever seen in action. **Junior** said:

"In order to gain the respect of your players in the locker room, you can't just perform on Saturday. You have to do it consistently during practices, meetings, and in the weight room."

Tad Sisler with NFL Hall of Famer Junior Seau

Source- Sisler Private Collection

Consistency is the key. Some people track their consistency by writing down their progress in a journal. Journaling is a powerful tool for tracking your progress and gaining insights into your health journey. In this book, I provide prompts to help you record your starting point—how you feel, any relevant measurements like weight or blood pressure, and your initial thoughts. As you go along, you'll also note down any "aha" moments when something clicks or you have a breakthrough. Celebrities like **Oprah Winfrey** swear by journaling for personal growth and reflection (*Oprah.com*). By keeping a journal, you can see how far you've come and what's working for you.

INTRO.2.2: LIFELONG INTEGRATION FRAMEWORK

Once you've got the hang of the initial actions, the book offers more advanced options to challenge yourself further. These are like levels in a game: tier-2 and tier-3 actions that increase in difficulty or complexity. For instance, if you started with walking, you might progress to jogging or adding strength training. **Hugh Jackman**, who transformed his physique for *Wolverine*, built his fitness step-by-step like this (*Men's Health*). These options keep you challenged and growing.

To help you turn these actions into lasting habits, the book includes monthly and quarterly checkpoints. These are times to review your progress, celebrate your successes, and adjust your goals if needed. It's similar to how businesses have regular performance reviews to ensure they're on track. By checking in regularly, you can make sure that your short-term efforts are leading to long-term changes.

In **Appendix A**, you'll find printable planners that allow you to track all your important health metrics in one place. You can record your lab results, heart rate variability (*HRV*), sleep quality, and your personal goals. Having everything together makes it easy to see connections and patterns. For example, you might notice that when you get better sleep, your HRV improves, indicating better recovery and less stress.

"It was character that got us out of bed, commitment that moved us into action, and discipline that enabled us to follow through." – Zig Ziglar

We know that everyone's different, so the book offers flexible options to suit various lifestyles and preferences. Whether you're vegetarian, following a keto diet, have specific faith-based dietary restrictions, or are just really busy, there are plans and menus that will work for you. **Beyoncé**, who's tried vegan and other diets for health and performance, shows how tailored plans can work (*Vogue*). These options make sure there's something for everyone, no matter your needs.

For the latest in weight management, check out my book **The Ultimate AI Diet - Consolidating the Best Diets Over the Last 100 Years: Achieve Fast, Lasting Weight Management with a Personalized, Modern Approach Backed by a Century of Proven Success** for a complete look into weight management from a modern perspective.

"I always have breakfast, say, scrambled egg whites, a vegetable smoothie, or whole-grain cereal with low-fat milk. For lunch and dinner, I eat a lot of fish and vegetables. And throughout the day, I try to stay hydrated."
– Beyoncé

Beyoncé Knowles
Credit – Wikimedia Commons

To make tracking your progress even easier, the book suggests using AI-powered tools and apps. These can automatically log your food intake, count your steps, and monitor your sleep patterns, providing you with feedback and suggestions. For instance, apps like *MyFitnessPal* can help with food logging, while wearables like *Fitbit* and *iWatch* track your activity and sleep. This automation saves you time and effort, allowing you to focus on making healthy choices (*MyFitnessPal, Fitbit*).

INTRO.2.3: TRACK → REFLECT → ADJUST LOOP

Before you start making changes, it's important to know where you stand. The book recommends getting baseline measurements for key health indicators:

- Your fasting insulin levels to check for insulin resistance.
- Your waist-to-height ratio as a measure of body fat distribution ..
- Your resting heart rate as an indicator of cardiovascular fitness
- Your DNA methylation age (DNAmAge) to estimate your biological age.

These metrics give you a clear starting point and help you track your progress over time. Tech entrepreneur **Bryan Johnson** is known for his extensive biohacking efforts, including tracking his biological age (*Bryan Johnson*).

"The discovery that we can program our existence and author our own lives has radically altered my life - transforming my identity, aspirations, and reason for existence." – Bryan Johnson

Bryan Johnson
Credit – Wikimedia Commons

Each week, take just two minutes to reflect on your progress. Note down what went well (your wins), what didn't go as planned (your misses), and what small adjustments you can make for the coming week. This quick review helps you stay accountable and continuously improve without taking up too much time. If you notice a decrease in your heart rate variability (HRV) or poorer sleep quality, it could be a sign that you're under too much stress or overtraining. The book provides strategies to adjust your routine accordingly, such as taking a rest day, practicing relaxation techniques, or reducing workout intensity. Even elite athletes like **Tom Brady** and **Novak Djokovic** use HRV monitoring through WHOOP and other recovery wearables to optimize their training and recovery, monitoring physiological indicators including HRV, sleep quality, and stress levels (*ESPN; The Upside*).

Every three months, you can retest your epigenetic age to see how your biological age is changing. This helps you assess whether your lifestyle changes are effectively slowing down the aging process. Companies like *Elysium Health* offer tests that measure your biological age based on DNA methylation patterns (*Elysium Health*).

Finally, this book encourages you to celebrate even the smallest improvements. Whether it's a 1% increase in your HRV, a slight decrease in your fasting insulin, or just feeling a bit more energetic, acknowledging these wins keeps you motivated. Remember, it's not about being perfect; it's about making consistent, small changes that add up over time.

Warren Buffett talks about the power of compounding in investing, and it applies to health too: small gains add up to big results (*Berkshire Hathaway*).

SECTION THREE: GROUND RULES/ EXPECTATIONS

INTRO.3.1: WHAT THIS BOOK IS AND IS NOT

This book is your trusty field guide, blending cutting-edge science with practical steps you can actually use to live longer and healthier. It's packed with research-backed strategies, explained in a way that's easy to grasp, so you can make smart choices for your body and mind without getting lost in technical jargon.

But don't expect a one-size-fits-all plan here. Everyone's body is different, so what works for one person might not for another. Always check with your doctor or healthcare team before diving into big lifestyle changes. This book is a helpful companion, not a replacement for professional medical advice.

Longevity is a marathon, not a sprint. It's about stacking up tiny victories—like eating a bit healthier or walking more—that add up over days, months, and years. Each small step you take builds toward a stronger, more vibrant life, much like how **Serena Williams** built her tennis career through consistent effort (*Forbes*).

"Luck has nothing to do with it, because I have spent many, many hours, countless hours, on the court working for my one moment in time, not knowing when it would come." – Serena Williams

Let's be real: there's no shortcut to living forever. This book steers clear of flashy promises like instant immortality or trendy detox cleanses you see on social media. Instead, it focuses on proven, science-based methods that deliver real results over time, keeping you grounded in what actually works. This book is a team effort. But I want to emphasize that we will all absolutely begin to experience age reversal and greater longevity quickly in the coming years due to unbelievable new discoveries around the corner.

You bring your curiosity and commitment, and it provides the roadmap to guide you. It's like having a coach who lays out the path, but you've got to lace up your shoes and walk it. Together, we'll work toward a healthier, longer life.

Speaking of having great coaches, I was fortunate to have a great friend and mentor in seven-time championship winning *NBA* basketball coach **Frank Hamblen**, who was **Phil Jackson's** right-hand-man with **Michael Jordan's** *Chicago Bulls* and **Kobe Bryant's** *Los Angeles Lakers*. **Frank** said:

"You just refuse to lose. True success is found in the relentless pursuit of excellence and the unwavering belief in your own potential."

Frank Hamblen and Tad Sisler

Source: Tad Sisler's Personal Collection

INTRO.3.2: STARTER TOOLKIT

To kick things off, grab a notebook or download a notes app to jot down your gratitude, health stats like weight or energy levels, and thoughts about your journey. Tracking these helps you spot patterns and stay motivated, just like keeping a diary helps you reflect on your goals. A basic wearable that tracks your steps and sleep is a great tool to understand your activity and rest patterns. No wearable? No problem—free phone apps can do the job just as well. These tools make it easy to see how your daily habits are shaping your health.

In your kitchen, a scale helps you measure portions for better eating habits, and a water bottle keeps hydration front and center. Step outside for some sunlight or fresh air, even if it's just on a balcony, to boost your mood and get a dose of vitamin D, which research links to better health (*Harvard Health*). A May 2025 study from Mass General Brigham (MGB) and the Medical College of Georgia reveals that taking Vitamin D supplements may protect against biological aging by slowing the shortening of telomeres (*Fox News*).

Want to level up? Consider optional tools like resistance bands for home workouts, a red-light panel for potential benefits like skin health, or blue-blocker glasses to cut down on screen glare before bed. These extras, explored later in the book, can add a little edge to your routine. Approach this with a beginner's curiosity, ready to try new things and tweak what doesn't work. Be open to testing, measuring, and adjusting, like a scientist experimenting in a lab. This flexible mindset will keep you moving forward, no matter where you start.

INTRO.3.3: YOUR FIRST MICRO-COMMITMENT

Assuming you're reading this book in an effort to change your life by adapting a new lifestyle, start with one small action that feels good: maybe write three things you're grateful for in a journal, take a 10-minute walk, or hit the hay a bit earlier.

Choose what excites you most—it's your first step toward a healthier you. Tell a friend or family member about your chosen action. Research suggests sharing your goal can boost your chances of sticking with it by up to 42% (*Journal of Behavioral Medicine*). Having someone cheer you on, like a teammate, makes all the difference. Remember the great **Lao Tzu** quote:

"The journey of a thousand miles begins with a single step."

Tonight, log your heart rate variability (*HRV*) with a wearable or app, or rate your mood on a 1-10 scale. Watch how these numbers shift as you try your new habit this week. Seeing progress, even small, is a huge motivator. Flip through the chapter intros to see what grabs you—maybe it's nutrition, exercise, or mental health. That spark of excitement is your launchpad, setting you up to dive into the book with enthusiasm. Ready to start? Take a deep breath, feel your pulse, and picture yourself at 90 (or even 120)—strong, vibrant, and full of life. Let that image fuel your journey as you take your first steps toward a healthier future.

EXTREMELY IMPORTANT – READ THIS BEFORE PROCEEDING

MEDICAL DISCLAIMER

The information presented in *The Unlimited Power of Your Mind and Body: How to Live Longer Naturally by Reprogramming Your Mind, Body, and Genes for Strength and Vitality* is provided solely for informational and educational purposes. It is not intended as a substitute for professional medical advice, diagnosis, or treatment provided by licensed healthcare professionals. The author and publisher are not medical professionals and do not offer medical advice. Readers should always consult their physician or another qualified healthcare provider before adopting new health regimens, including dietary adjustments, exercise routines, supplements, mindfulness practices, or any other strategies described in this book.

The methods, practices, and recommendations provided in this book may not be appropriate or safe for everyone. Individual effectiveness and safety may vary widely depending on personal health conditions, genetics, lifestyle, and other unique factors. Some strategies, techniques, or concepts described herein may be experimental or still under scientific investigation, and their availability and validity may be limited or subject to change as further research emerges.

No warranties or guarantees are provided regarding the accuracy, completeness, or appropriateness of any information in this book. The reader assumes full responsibility for any actions taken or decisions made based on the content of this book. The author and publisher expressly disclaim liability for any injury, loss, or harm resulting directly or indirectly from the use, misuse, or misunderstanding of information provided in this book.

LEGAL DISCLAIMER

The Unlimited Power of Your Mind and Body: How to Live Longer Naturally by Reprogramming Your Mind, Body, and Genes for Strength and Vitality is published strictly for informational and educational purposes. It does not provide professional medical, legal, financial, psychological, or other professional advice. Although efforts have been made to ensure the accuracy and reliability of the content, the author and publisher make no representations or warranties about its accuracy, completeness, or suitability for individual circumstances.

Readers assume sole responsibility for any decisions made or actions taken based on the book's content. By engaging with this material, the reader agrees to hold harmless the author, publisher, and any affiliates from any claims, demands, liabilities, or damages that may arise directly or indirectly from the application or misuse of any information within this book.

This book does not establish any professional-client relationship between the reader and the author or publisher. Readers are encouraged to seek guidance from qualified professionals for specific concerns or decisions. Mention of specific products, techniques, or practices does not constitute endorsement or warranty by the author or publisher.

The information provided, including references to supplements, therapies, genetic interventions, mindfulness practices, or emerging research, is subject to change with evolving scientific discoveries and clinical experiences. The author and publisher expressly disclaim any responsibility for errors or omissions and are not liable for any consequences arising from the reliance upon or application of information presented in this book.

****

Thoughts shape biology, but biology still ages... What exactly wears us down? Now, let's discover the hidden cellular clocks—and how to start winding them back.

CHAPTER ONE
THE BIOLOGY OF AGING IN PLAIN ENGLISH

Here's a statistic: **Every minute, 50 million of your cells die—yet you wake up fine tomorrow. Why?** How does that amazing healing power work within you? In the 1980s movie *E.T.*, the cute little alien merely touches the child's wound with his glowing finger, and it perfectly heals instantly. Are we approaching technology that can heal us in the same way—or better yet, actually reverse our aging process?

Aging is a natural process that affects us all, but what exactly happens inside our bodies as we get older? In this chapter, we'll explore the biological mechanisms that drive aging, explained in simple terms. From the tiny structures in our cells to the ways we can measure and manage our age, you'll gain a clear understanding of how aging works and what you can do to influence it positively.

SECTION ONE: CELLULAR TIMEKEEPERS

To understand aging, we need to look at what's happening inside our cells. There are several key players that act like timekeepers, determining how our cells age over time. These include telomeres, senescent cells, and epigenetic clocks. I delve much deeper into all of these amazing players in my book **Stay Healthy, Stay Youthful: The Science of Living to 150: Cutting-Edge Longevity Science for Reversing Aging and Living a Healthier, Longer Life.**

1.1.1: TELOMERE TICKERS

Your DNA has protective caps called telomeres, like the tips on shoelaces, that get shorter every time a cell divides. Stress and smoking can speed up this process, but meditation might slow it down, as shown in a University of California, San Francisco study (*UCSF Meditation Study*). An enzyme called telomerase can rebuild these caps, and it's boosted by exercise and omega-3s from foods like salmon. You can even check your telomere length with home test kits.

In Okinawa, Japan, centenarians often have longer telomeres; a 102-year-old farmer's long life, tied to his active lifestyle and healthy diet, reflects this (*Okinawa Centenarian Study*).

1.1.2: ZOMBIE (SENESCENT) CELLS

As we age, some cells stop dividing but don't die, turning into "zombie" cells that release harmful substances, causing inflammation in places like joints and skin. Fasting, high-intensity interval training (*HIIT*), and fisetin from strawberries can help clear them out. A mouse study found that removing 30% of these cells extended life by 25% (*Senescent Cell Study*). If your joints ache or wounds heal slowly, you might have too many of these cells. Scientists are also testing drugs called senolytics to target them.

1.1.3: EPIGENETIC CLOCKS

Your genes have chemical tags, like sticky notes, that control how they work, and these change with age. The **Horvath Clock** measures 353 of these tags to estimate your biological age—how old your body really is. A small study showed lifestyle changes, like better diet and exercise, can rewind this clock by about two years in eight weeks (*Epigenetic Age Reversal*).

Stress, though, can mess up these tags, raising disease risks. Many people, including well-known figures, have explored biological age testing and lifestyle changes to look and feel younger. For example, a recent case series showed that women could reduce their biological age by an average of 4.6 years through an 8-week program involving a plant-based diet and other interventions (*Biological Age Case Series*).

SECTION TWO: DAMAGE VS. DEFENSE

While our cells have mechanisms to keep time, they also face constant threats from damage. Let's explore how our bodies defend against these threats and what we can do to support them.

1.2.1: OXIDATIVE STRESS AND FREE RADICALS

Every day, your body produces free radicals—unstable molecules that can damage cells—as a result of normal processes like breathing and exercise, as well as from external factors like pollution. To combat this, antioxidants from whole foods, such as cold-pressed extra virgin olive oil and berries, are more effective than high-dose supplements because they work together synergistically. Interestingly, a small amount of these reactive oxygen species (ROS) can actually be beneficial through a process called hormesis, which strengthens your cells, but too much leads to tissue damage. For example, smokers show higher levels of lipid peroxidation, a marker of cell damage (*Smokers' Lipid Peroxidation*). Eating colorful fruits and veggies, like those in our already mentioned friend **Tom Brady's** diet, helps keep this damage in check. As an NFL quarterback chosen in late rounds of the draft, **Brady** knew early on that he needed to do whatever it took as the underdog to become the greatest of all time. Diet and exercise were on top of his list. While he was still playing, he said:

"I was the kid that was the 199th pick that never had the body for it. People didn't think I'd play one year in the NFL, and now I'm going on my 17th year."

Tom Brady

Credit – Wikimedia Commons

1.2.2: MITOCHONDRIAL DECLINE

Mitochondria, often called the powerhouses of the cell, generate the energy your body needs. As we age, they become less efficient, which can leave you feeling tired or mentally foggy. Fortunately, there are ways to support them. Red and near-infrared light therapy, in the 660–850 nm range, can enhance their function by boosting an enzyme called cytochrome c oxidase. Supplements like CoQ10, PQQ, and creatine can also help increase ATP production, the molecule that provides energy. I go into great detail on these and other great supplements in my book **Vitamins, Supplements, and Herbs for Health and Longevity: Boost Your Immunity, Increase Energy, and Feel Younger in Minutes a Day.** Additionally, ketone esters offer an alternative fuel source that can reduce the strain on mitochondria. Elite athletes, like ultra-runner **Scott Jurek**, who engage in zone-2 training—a moderate intensity workout—often have significantly higher mitochondrial density, which helps them sustain energy over long distances.

1.2.3 DNA Repair Crews

Your cells have built-in repair mechanisms to fix DNA damage, but these rely on a molecule called NAD+, which decreases as you age. Enzymes like PARP need NAD+ to function properly. You can boost your NAD+ levels through fasting and high-intensity interval training (HIIT), which also activate sirtuins—proteins linked to longevity. Getting enough sleep, ideally 7 to 9 hours, provides a nightly window for DNA repair. On the flip side, drinking too much alcohol can deplete these repair proteins quickly. There's also emerging research on rapamycin, a drug that shows promise in extending lifespan in animal studies, but it's important to discuss this with your doctor before considering it.

SECTION THREE: MEASURING AND MANAGING YOUR AGE

Understanding the science is one thing, but how do you apply it to your own life? In this section, we'll look at tools and strategies to measure your biological age and manage it effectively.

1.3.1: LAB AND WEARABLE DASHBOARD

To keep tabs on your aging process, it's helpful to monitor certain biomarkers. Key ones include fasting insulin to check for insulin resistance, hs-CRP for inflammation, ApoB for heart disease risk, DNAmAge for your biological age, and heart rate variability (HRV) for stress and recovery. You can also use a simple grip strength dynamometer, as studies show that grip strength is correlated with overall mortality risk (*Grip Strength Mortality*). Wearable devices can track your HRV; if it drops below 40 milliseconds, it might be time to take a break and recover. Getting lab tests every three months helps you see trends over time, and remember to keep your health data secure by encrypting it.

1.3.2: 72-HOUR JUMP-START

Ready to start making changes? Here's a quick plan to jump-start your journey.

- First, order a home epigenetic test to find out your biological age, and take a selfie to visually track your progress.
- After each meal, go for a 10-minute brisk walk; studies show this can lower your blood sugar significantly (*Post-Meal Walking*).
- Swap out one ultra-processed snack for a handful of raw nuts—nuts are packed with healthy fats and nutrients.
- Try going to bed 30 minutes earlier than usual and track your morning heart rate to see how rest affects you.
- Finally, write down in one sentence what you want your life to look like at age 90. This vision will keep you motivated, just like how **Oprah Winfrey** uses vision boards to focus on her goals (*Oprah's Vision Boards*).

1.3.3: LIFELONG INTEGRATION

To make these changes stick, integrate them into your lifestyle over the long term. Consider doing a 24-hour fast once a month, followed by a sauna session, to help your cells clean out old components. Aim for at least 150 minutes of moderate-intensity cardio, like brisk walking or cycling, and two strength training sessions each week—think of it like **Dwayne "The Rock" Johnson's** workout routine, which keeps him in top shape (*Dwayne Johnson's Workout*). Diversify your diet by eating 30 different plant species weekly to boost your gut microbiome's diversity; a high *Shannon* score indicates a healthy gut.

Every three months, take a moment to reflect on which health markers have improved the most and why. And don't forget to celebrate every small victory—whether it's better sleep or more energy—because these micro-wins compound over time, leading to significant health improvements.

Now that you know what ticks and rusts inside every cell, let's explore the on-switch hiding between your ears. In the next chapter, we'll see how **a single thought can change your biochemistry in seconds.**

CHAPTER TWO
THOUGHT PATTERNS REWIRING BIOLOGY

"Thoughts are things," and powerful things at that, when they are mixed with definiteness of purpose, persistence, and a burning desire..."
- Napoleon Hill

This chapter is exciting to me! Just think: Change your thoughts, change your cytokines. That's right—what you think can shift the very chemicals in your body that control inflammation, immunity, and even aging. Scientists have found that negative thoughts can spike harmful markers like IL-6, while a simple gratitude practice can boost protective antibodies in minutes. Your mind isn't just a control center for emotions; it's a switchboard for your biology, and this chapter will show you how to flip the right switches to feel better, live longer, and thrive. Changing your thoughts can change your life. My friend, legendary *NBA* legend **Earvin "Magic" Johnson** had his share of disappointments in his career. **Magic** was a great example of someone who believed that a great attitude and hard work could override other people's perception of him. **Magic** said:

"If somebody says no to you, or if you get cut, Michael Jordan was cut his first year, but he came back and he was the best ever. That is what you have to have. The attitude that I'm going to show everybody, I'm going to work hard to get better and better."

Tad Sisler on the L.A. Lakers Bench with Earvin "Magic" Johnson
Source – Sisler Private Collection

SECTION ONE: MIND-BODY CIRCUITRY

Your brain and body are in constant conversation, and the way you think can change how your body works. New developments in brain research include **Elon Musk's** *Neuralink,* allowing people with ALS who can no longer talk to 'think' their words into a device that speaks for them. Let's explore the science behind this and other mind-body connections and how you can use it to your advantage. **Bob Proctor** said, ***"What you think about, you bring about."*** This is how it works.

2.1.1: PSYCHONEURO-IMMUNOLOGY 101

Your thoughts kick off a chain reaction in your body, starting with the hypothalamus in your brain, which triggers a hormone cascade affecting everything from stress to immunity. Negative thinking, like endlessly worrying about a deadline, can raise inflammation markers like IL-6 and TNF-α, which over time can harm your health (*Negative Thoughts and Inflammation*). On the flip side, visualizing positive outcomes—like picturing a great day—can boost protective antibodies like IgA in as little as 10 minutes. A study even showed that placebo knee surgery worked just as well as the real thing for pain relief, proving the mind's power (*Placebo Knee Surgery Study*). For example, an IBS patient saw their symptoms drop significantly after a course of cognitive behavioral therapy, showing how mental shifts can lead to physical healing.

2.1.2: BRAIN-HEART COHERENCE

Heart rate variability (HRV) gives you a peek into your body's balance between stress and calm, and you can improve it with a technique called brain-heart coherence. Breathing at a pace of 5.5 breaths per minute—about a 5-second inhale and exhale—puts you in peak coherence, dropping stress hormone cortisol by 23% (*HeartMath Coherence Study*). Biofeedback apps make this practice fun by gamifying it, letting you see your coherence improve in real time. Doing this with a group, like a meditation circle, can amplify the effect three times, and practicing before a big meeting can sharpen your decision-making clarity, keeping you cool under pressure.

2.1.3: REFRAMING THE STORY

The way you think about events shapes how they affect you, and the ABC model breaks it down: an event leads to a belief, which leads to a consequence. Adding "yet" to your thoughts—like ***"I'm not good at this yet"***—shifts you from a fixed to a growth mindset, opening up possibilities. Keeping an evidence log, where you write down facts that challenge negative thoughts, helps counter distortions like ***"I always fail."*** Humor can also stop your brain's "doom loops," as the amygdala calms down when you laugh.

A firefighter used this by visualizing successfully lifting debris during training, which later helped him save a crewmate in a real emergency, showing how mental reframing can lead to life-changing actions.

When I was in sixth grade, teachers at my elementary school divided students into three groups. The first group played basketball every day. The second group was taught to visualize playing basketball, going through the mental motions of dribbling and shooting baskets. The third group did nothing at all. At the end of the study, the group that visualized was almost as good as the group who actively participated in the game, illustrating that when you think about an event as if you're doing it, your mind reframes the experience into actuality.

SECTION TWO: EVERYDAY MENTAL TOOLS

Now that you know how thoughts affect your body, let's look at simple tools you can use every day to rewire your mind for better health. You can read more about these tools in my book **The Science of Positive Thinking: How Mindset, Daily Habits, and Emotional Well-being Can Add Years to Your Life.**

2.2.1: FIVE-MINUTE MEDITATION MENU

Meditation doesn't have to be complicated—just five minutes can make a difference, and you've got options to choose from. Try box breathing: inhale for 4 seconds, hold for 4, exhale for 4, and hold for 4, repeating the cycle to calm your mind. "Noting" mindfulness means labeling thoughts as they pop up—like "thinking" or "worrying"—then letting them go without judgment. Loving-kindness meditation involves wishing well for yourself and others, starting with ***"May I be happy"*** and extending it outward. A body scan, where you focus on sensations from your toes to your scalp, helps you relax. If focusing on your breath feels tricky, anchor your attention on ambient sounds, like birds chirping or a fan humming, to stay present.

2.2.2: GRATITUDE AND OPTIMISM LOOPS

Gratitude and optimism can transform your mindset and even your biology. Writing down three good things each day—like a sunny walk or a kind word—can boost your happiness by 25% in just one week (*Three Good Things Study*). A study from the University of Pennsylvania found that writing a gratitude letter to someone can lower depression symptoms for up to three months (*Gratitude Letter Study*). Visualizing an optimistic future has been shown to lengthen telomeres, those DNA caps that protect against aging, in small studies. Sharing gratitude publicly, like posting a thank-you on social media, amplifies the effect through social connection. A nurse who started sending daily gratitude texts to her team saw morale on her ward skyrocket, proving how contagious positivity can be.

2.2.3: CBT MICRO-EXERCISES

Cognitive Behavioral Therapy (CBT) isn't just for therapy sessions—you can use mini-exercises to shift your thinking anytime. Catch Automatic Negative Thoughts (ANTs) as they happen, like ***"I'm terrible at this,"*** and question their truth. Swap ***"I have to"*** for ***"I choose to,"*** turning chores into choices that feel empowering. Rate your worries on a catastrophe scale from 1 to 10—most things aren't a 10, which puts them in perspective. Re-script worst-case scenarios in your head by adding a funny soundtrack, like cartoon music, to lighten the mood. Finally, use a weekly thought audit worksheet to track and challenge negative patterns, helping you build a more balanced mindset over time.

My friend, former **President Gerald R. Ford** was a star football player in college. He wasn't so adept at golf, and he was known to hit many errant golf balls in celebrity golf tournaments. Fortunately, he had a good sense of humor and he was able to reframe the experience into a positive mindset when he said:

"I know I am getting better at golf because I am hitting fewer spectators!"

President Gerald R. Ford and Tad Sisler

Source – Sisler Private Collection

SECTION THREE: FROM THOUGHT TO HABIT

You've got the tools—now let's turn them into habits that stick. This section gives you a quick start, tracking methods, and long-term strategies to make positive thinking second nature.

3.3.1: 72-HOUR JUMP-START

Let's get started with a simple three-day plan to rewire your thoughts.

• Each morning, spend two minutes on slow breathing—like a 5-second inhale and exhale—followed by writing down one thing you're grateful for.

• At mid-day, set a phone reminder to check your posture and smile, which instantly lifts your mood.

• In the evening, jot down three good things from your day and do a 90-second body scan, noticing any tension from toes to head.

• Rate your mood on a 1-to-10 scale and log your HRV each night to track how these habits affect you.

• Share this plan with a friend—having an accountability buddy makes you more likely to stick with it, just like workout partners keep you motivated.

3.3.2: TRACKING PROGRESS

Keeping track of your progress helps you see what's working.

• Each week, graph your HRV trends against your mood to spot patterns—like how a good night's sleep might boost both.

• Tally your negative thought distortions, aiming to reduce them over time, and use a biofeedback app to earn "star streaks" for minutes spent in coherence, making it fun to stay consistent.

• Celebrate your wins with non-food rewards, like a walk in the park or a favorite playlist, to keep the momentum going.

• If you notice dips in mood or HRV, troubleshoot by checking your sleep, caffeine intake, or if you've been doom-scrolling the news too much.

3.3.3: LIFELONG INTEGRATION

To make these mental habits a permanent part of your life, build on them gradually.

• By week four, upgrade to a 10-minute meditation session to deepen your practice, like **Ellen DeGeneres**, who's known for her mindfulness routines to stay grounded (*Ellen's Mindfulness*).

• Plan a quarterly silent retreat or digital detox weekend to reset your mind. Teach one of these mindset tools—like gratitude journaling—to a friend; teaching helps you master it.

• Stack gratitude with a daily habit, like saying thanks while sipping your morning coffee, to make it stick.

• Write a letter to your future self about your health goals, keep it in your nightstand, and reread it monthly to stay inspired.

Your mindset switch is on. But emotion is rocket fuel—or rocket crash. Next up: learn to steer feelings so they build health instead of eroding it.

CHAPTER THREE
EMOTIONAL MASTERY AND EUSTRESS

Be happy and live longer. Chronic pessimists die 7.2 years earlier than optimists—even with identical diets and exercise, according to a long-term study tracking emotional outlook and health outcomes (*Pessimism and Longevity Study*). That's a stark reminder of how much your emotions matter. In this chapter, we'll unpack the biology behind your feelings, share hands-on tools to harness them for good, and build daily habits to keep your emotional rocket fuel burning bright without crashing.

SECTION ONE: BIOLOGY OF EMOTION

Emotions aren't just feelings—they're chemical signals that shape your health. Your attitudes and emotions can shape your successes or failures. Let's break down how they work in your body and brain, and how to turn them into allies for longevity.

3.1.1: THE 90-SECOND RULE

Emotions are fast but fleeting—neuroscientist **Jill Bolte Taylor** says their chemical surge in your body lasts about 90 seconds, and anything longer is you replaying the feeling in your mind. Naming what you feel, like saying ***"I'm angry,"*** shifts brain activity from the emotional amygdala to the logical pre-frontal cortex, calming you down. A deep nasal inhale through your left nostril can activate your calming parasympathetic system, and a Harvard study found that labeling anger reduces amygdala activity by 30% (*Harvard Emotion Labeling Study*). A CEO used this trick to defuse tension in a heated boardroom meeting, taking 90 seconds to name his frustration and breathe, bringing the room back to focus.

3.1.2: EUSTRESS VS. DISTRESS

Believe it or not, as much as sitting on a beach with your toes in the sand for the rest of your life sounds great at any given moment, you don't want zero stress in your life. Stress isn't always the enemy—it follows a U-shaped curve where too little means boredom, too much causes damage, but the middle is where growth happens. A short cortisol spike from stress can be helpful, but a constant high level leads to belly fat and telomere shortening, speeding up aging. Deliberate micro-stresses, like a cold shower or high-intensity interval training (HIIT), train your resilience by giving you controlled pulses of stress.

Reframing stress as a "challenge" can improve your DHEA-to-cortisol ratio, boosting your body's ability to handle it, as shown in Stanford research (*Stanford Stress Reframing*). A 30-year study found that kids who viewed stress as helpful lived longer, proving the power of mindset in handling life's pressures.

Anger, jealousy, envy... all these negative emotions cause stress. They're **acids that eat their own container**. Learning to let go and go with the flow will keep you healthier in the long run. My friend, legendary CNN broadcaster **Larry King** had a modicum of stress in his daily life but he knew how to handle it. He said you wouldn't want to be with him during a traffic jam because he outwardly expressed his anger, which actually worked in his favor along with the 90-second rule. But after a couple heart attacks, he also learned how to let go by not resenting others or holding a grudge. **Larry** said:

"I don't wish anybody ill, I really don't. I've got a lot of faults, but I was never jealous or envious or... it's a waste of time."

Larry King and Tad Sisler
Source – Sisler Private Collection

3.1.3: FLOW STATES

Larry King learned to go with the flow in part because he immersed himself and the viewer into his in-depth interviews. Flow states happen when you're so absorbed in an activity that you lose track of time—it's that sweet spot where the challenge is just 4% beyond your comfort zone, balancing skill and difficulty. In flow, your brain floods with dopamine and norepinephrine, sharpening focus and even muting pain, which is why two hours can feel like 20 minutes. You'll know you're in flow when you've got clear goals, immediate feedback—like seeing progress in real time—and a touch of risk to keep things exciting. A coder once pulled a 14-hour bug-fixing sprint in seamless flow, solving a complex problem without noticing the hours slip by, showing how flow can supercharge productivity and satisfaction.

In my ten year career as a music developer for Yamaha, much of what I did was tedious, making every note in a programmed song perfect on multiple tracks. Losing myself into the flow and enjoying the process made the job seem easier over time.

Going with the flow is like dancing with life's rhythm, balancing action and patience to boost your health and vitality. It means listening to your body—knowing when to push with exercise or rest for recovery—without forcing things. Remember the saying, ***"The second mouse gets the cheese."*** Be cautious; don't rush into health fads that could backfire. Yet, stop and smell the roses to ease stress and enjoy life's moments. Let's explore a couple of conflicting statements. You may have heard the saying, ***"He who hesitates is often left behind."*** This concept urges timely health steps. You've probably also heard that ***"Good things come to those who wait."*** This idea values the idea of allowing your body to heal or work through problems. You can embrace both concepts at the same time! Not procrastinating on things that can improve your health is great. It's also great to give yourself a chance to catch up and move forward. This mindset reduces stress hormones, enhancing your body's vitality, as a 2025 study shows mindfulness lowers cortisol by 20%.

This approach reprograms your mind and body for a longer life. By staying adaptable, you lower inflammation and support gene expression for health, per 2025 *Nature Review Genetics* research. Practicing mindfulness with apps like Headspace or taking daily walks helps you adapt to life's changes, reducing stress and boosting your immunity. Joining support groups keeps you connected. Going with the flow means trusting your body's signals while taking smart steps to create a healthier, stronger you. Start today to live longer and thrive!

3.1.4: COLLECTIVE COHERENCE AND GLOBAL FIELD EXPERIMENTS

We humans are connected in ways we don't quite yet understand. For instance, your emotions don't just affect you—they can ripple out to others in surprising ways. The Princeton Global Consciousness Project found that random-number generators act differently during big emotional events—like global celebrations or tragedies—suggesting our collective feelings might influence the world around us (*Princeton Global Consciousness Project*). This is really cool because a particular outcome sequence normally will contain some patterns detectable in hindsight but impossible to foresee. If large groups of humans can alter these with their feelings, it proves that collective consciousness exists within all humans. So, we're all connected on some "psychic" level. A HeartMath pilot program called "Heart-brain Global Coherence" showed that when people sync their heart rhythms in a group, each person's immune-boosting IgA antibodies rise by 12%, making everyone stronger together (*HeartMath Coherence Study*).

Another study at the College of New Jersey used fMRI to show that eight meditators focusing together had their brain's thinking and attention areas sync up, creating a powerful group connection. You can try this yourself—gather a few friends for a small "collective-focus" session, maybe meditating or setting a positive intention together, and track how your HRV or mood score changes to see the group magic in action. On a personal note, with probably too much information here, my four older sisters found that their menstrual cycles all synced up together when they lived in the same house as teenagers. So, there was about one week per month I did not want to be around any of them!

SECTION TWO: HANDS-ON EMOTIONAL TOOLS

You don't need a lab to manage your emotions—here are simple tools you can use anywhere to turn stress into strength and keep your feelings working for you.

3.2.1: EFT TAPPING

Emotional Freedom Technique (EFT) tapping combines meridian-style tapping with a cognitive statement to calm your body fast—studies show it can lower cortisol by 37% (*EFT Tapping Cortisol Study*).

• Tap in sequence: start on the side of your hand, then move to your eyebrow, side of your eye, under your eye, and so on, while saying, ***"Even though I feel emotion, I deeply accept myself."***

It's easy to follow with a 5-minute mobile app walkthrough. My sister's husband, a veteran with PTSD, used this tapping routine to quell a flashback, finding calm in minutes by focusing on his breath and the rhythm of the taps.

3.2.2: HEART-BRAIN COHERENCE

Technically speaking, heart-brain coherence uses HRV biofeedback to get your heart rhythm into a smooth sine wave at 0.1 Hz, which you can achieve by breathing at 5.5 breaths per minute—a 5-second inhale and exhale. While breathing, picture something you're grateful for to spike your IgA antibodies, which protect your immune system. Apps like *Inner Balance* and *Elite HRV* turn this into a game, tracking your coherence streaks to keep you motivated. Group sessions can amplify the effect, and corporate teams using this have seen sick days drop by 18%. Just two minutes of this practice before a meeting can sharpen your decision-making, helping you stay clear and focused.

"The first step toward change is awareness. The second step is acceptance." – Nathaniel Branden

3.2.3: LAUGHTER MEDICINE

Just like author **Norman Cousins** realized as he laughed his way back to good health, laughter isn't just fun—it's medicine. A 10-minute genuine laugh session can boost your natural killer (NK) cell activity for two hours, strengthening your immune system (*Laughter and NK Cells*). It also releases endorphins and oxytocin, relaxing your blood vessels and lowering blood pressure. Laughter yoga starts with fake giggles that turn real, thanks to mirror neurons making you catch the joy. Watching a comedy clip at lunch can lower cortisol, and the calm lasts until dinner. An oncology ward's "Friday Funnies" sessions—where patients watched funny videos together—cut pain medication requests in half, showing laughter's power to heal. Imagine that.

BIRDS CAN SENSE IF YOU ARE A FRIEND OR ENEMY

Your smile does more than brighten your day—it can even calm the birds around you! Studies suggest birds, like some animals, may sense human emotions through facial expressions, reacting with less fear to a smiling face than a stern one, like a scarecrow's frown. This illustrates how your positive emotions spread outward, influencing not only your health but also the world around you while reprogramming your mind and body for vitality. A positive attitude, expressed through a smile, reduces stress hormones like cortisol, boosting your immune system and heart health, per a 2025 study in the *Journal of Psychosomatic Research.* By smiling more, you create a healthier you and a calmer environment, fostering strength and longevity.

This connection extends to how your emotions shape your interactions. Just as birds respond to your smile, your positive mindset can inspire others, creating a supportive community that enhances well-being. Practicing daily smiles, gratitude, or kind gestures rewires your brain for optimism, supporting gene expression for vitality, as shown in the 2025 *Nature Reviews Genetics* research. Embrace your smile as a tool to reprogram your health and connect with the universe, living a longer and stronger life.

SECTION THREE: DAILY PRACTICES AND INTEGRATION

Let's build emotional mastery into your daily routine with morning checks, evening releases, and a ritual to carry these habits forward.

3.3.1: MORNING EMOTION CHECK

Start your day by tuning into your emotions—it only takes a minute. Rate your mood from 1 to 10 on your phone and jot down what's driving it, like a big meeting or a good night's sleep. Take three deep diaphragm breaths while naming your state out loud, like ***"I feel anxious."*** Set a "feel target" for the day—maybe you want to feel curious or energized—and choose one action to nurture that, like listening to an upbeat song. Boost the effect by stepping into sunlight and moving, like a quick walk, which multiplies the emotional lift.

3.3.2: EVENING EMOTIONAL RELEASE

Unwind at night with a simple routine to let go of the day's stress. Write a two-sentence journal entry: one sentence about what felt heavy, like a tough conversation, and another about the lesson you learned, like ***"I need to set better boundaries."*** Spend five minutes on a gentle yin stretch to open your fascia and signal your body to relax. Sip a tea with 3 grams of glycine to lower your core temperature for better sleep and avoid news or screens in the last hour to keep your amygdala calm. End by recalling a moment of gratitude, rewiring your hippocampus to focus on the positive, much like **Oprah Winfrey** does with her nightly gratitude practice (*Oprah's Gratitude Practice*).

3.3.3: SEGUE RITUAL

Set yourself up for tomorrow with a quick ritual to keep your emotions on track. Pick a "keystone emotion" you want to feel, like confidence or calm, and spend 30 seconds visualizing yourself succeeding with that feeling—like nailing a presentation with a smile. Place a sticky note on your mirror with your chosen emotion as a morning cue. Flash a quick smile to prime your mirror neurons for positivity, just like **Will Smith** does to start his day with optimism (*Will Smith's Morning Routine*). Tomorrow we'll turn that emotional rocket fuel into raw physical power.

"Throughout life people will make you mad, disrespect you and treat you bad. Let God deal with the things they do, cause hate in your heart will consume you too." – Will Smith

Will Smith

Credit – Flickr / creativecommons.org

CHAPTER FOUR
MOVE LIKE YOU MEAN IT!

Muscle after 40 is your most valuable retirement account—lose 1% a year or invest just 10 minutes a day to keep it strong and thriving. Research shows that staying active as you age can add years to your life, and it's not about marathon gym sessions; small, smart movements can make a big difference. In this chapter, we'll explore how to build strength that could save a life, boost your endurance without rigid diets, and weave movement into your day so it feels effortless.

My good friend, *Major League Baseball* Hall of Fame Pitcher **Trevor Hoffman** understands that it's a daily effort. He and I go to the same gym in San Diego, and he's dedicated to keeping his body in shape well after his career as a pitcher. **Trevor** said:

"I pride myself on being consistent. So, yes, it definitely means something, even though I don't set goals."

Tad Sisler with Trevor Hoffman

Source – Sisler Private Collection

SECTION ONE: BUILDING FUNCTIONAL STRENGTH

Strength isn't just about looking good—it's about being ready for life's challenges, from lifting groceries to handling emergencies. Let's break down how to train your body to be strong and resilient.

4.1.1: EMERGENCY-STRENGTH CIRCUITRY

Your body has a built-in governor that limits your muscle output to about 40% to keep you safe, but you can train it to tap into more when it counts. Try an isometric "door-frame push"—press as hard as you can against a door frame for 10 seconds—to teach your nervous system to recruit more muscle fibers. Using the Valsalva maneuver (holding your breath) while bracing your diaphragm creates spinal stability, making you stronger and safer.

Doing three 10-second max pushes a day can boost your neural drive by 15% in just two weeks (*Isometric Training Study*). A teacher once used this strength to lift a vending machine off a student during an accident, and a post-incident MRI study showed her nervous system had adapted to access more power under stress.

4.1.2: MUSCLE PROTEIN SYNTHESIS (MPS)

Building muscle after 40 hinges on muscle protein synthesis, or MPS, which you can kickstart by hitting a leucine threshold of about 2.7 grams per meal—which is about the amount of a serving of chicken or tofu. Aim for 0.8 to 1 gram of protein per kilogram of body weight (or about 0.5 grams per pound) daily to balance longevity and muscle growth. Resistance bands can get you 80% of the MPS you'd get from free weights, making them a great at-home option. After a workout, 25 grams of whey or a plant-based protein blend maximizes repair, and don't mess with sleep—it's when your growth hormone peaks to rebuild muscle, so skip late-night caffeine (*Protein and MPS Study*). I'll delve deeper into the vast benefits of sleep in Chapter 6.

4.1.3: MITOCHONDRIAL TURBO MODE

Your fast-twitch muscle fibers are packed with mitochondria, the energy powerhouses, and training them can put your energy production into turbo mode. Just 10 minutes of red-light therapy at 660 nm can prime your mitochondria's cytochrome-c, boosting ATP production by 14%. Creatine, at 5 grams a day, fills your muscle's energy reserves, while short, all-out sprints can double a gene called PGC-1α that sparks mitochondrial growth. A 70-year-old grandad saw his sprint power soar after combining red light and creatine, proving you're never too old to get stronger—kind of like how **Tony Horton** keeps rocking fitness in his 60s with smart training (*Red Light and Mitochondria*).

SECTION TWO: ENDURANCE AND METABOLIC FLEXIBILITY

Endurance isn't just for athletes—it's about having the stamina to live your best life, and it starts with teaching your body to use energy efficiently. Let's look at how to build lasting energy and flexibility.

4.2.1: CENTRAL-GOVERNOR THEORY

Your brain has a built-in governor that slows you down before you're truly exhausted to protect you, but you can push past it with the right mindset. Self-talk like ***"I'm stronger each step"*** can extend your time-to-exhaustion by 15%, tricking your brain into letting you go further. Olympic runner **Usain Bolt** famously used self-talk during his final sprints, visualizing victory to power through and win gold (*Usain Bolt's Self-Talk*).

Caffeine, at about 3 mg per kilogram of body weight, blocks the "brake" signal adenosine, giving you a mental edge. Playing music at 140 beats per minute distracts your governor, keeping you moving (think *"Toxic"* by **Brittney Spears** or *"Eye Of the Tiger"* by **Survivor**).

BATAAN DEATH MARCH AND SURVIVOR'S BRAINS

My father was a veteran of World War II. He had many stories of his own of seeing his shipmates blown apart during terrible battles. His own best friend died in his arms in one of those battles. No WWII story was more devastating than that of the brutal Bataan Death March. Survivors faced an ordeal that pushed them way beyond ordinary human limits—marching more than 60 grueling miles without enough food, water, or rest. As many of their own dropped and died from exhaustion in front of their very eyes, these survivors later recalled hitting that same wall of exhaustion so intense it felt impossible to take another step. Yet, somehow, they found a way to push through. Without realizing it, they managed to bypass their brain's built-in governor, that safety switch designed to protect us from pushing our bodies too far. By drawing on sheer determination and powerful mental resilience, they tricked their own brains into letting them keep going past what should have been their absolute limit.

The psychological tricks these soldiers naturally relied on—like breaking the endless march into small, achievable chunks, drawing strength from cherished memories, and focusing their minds on the camaraderie of fellow prisoners—allowed them to temporarily silence their bodies' desperate pleas to stop. Their remarkable stories reveal an incredible truth: when survival is at stake, the human mind can override our biological safeguards, tapping hidden reserves of strength and stamina. These survivors didn't make it through simply by luck; they survived by rewriting the rules of their own mental and physical endurance.

Bataan Death March

Credit – NARA & DVIDS Public Domain Archive

4.2.2: FAT ADAPTATION WITHOUT KETO DOGMA

You can train your body to burn fat efficiently without going full keto by focusing on Zone-2 training, where you exercise at a pace that burns up to 70% fat—think a brisk walk where you can still talk. A two-day low-carb primer before a big event can improve your ability to switch between carbs and fat for fuel, making you more efficient. Using a continuous glucose monitor (CGM) shows a flatter blood sugar curve after this adaptation, meaning fewer energy crashes. Ketone esters can give your brain a quick boost during long events, and refeeding carbs afterward helps balance your thyroid, keeping your metabolism humming (*Zone-2 Fat Burning*).

4.2.3: LACTATE AS FUEL, NOT FOE

Lactate isn't the bad guy it's made out to be—unless you let it build up. It's actually a fuel your body uses, shuttling energy between muscles and your brain through the "lactate shuttle."

High-intensity intervals teach your cells to recycle lactate faster, turning it into energy instead of letting it build up. Drinking beet juice, packed with nitrates, can increase lactate clearance by 18%, helping you recover quicker. Cold-water immersion after a workout boosts a transporter called MCT-1, which helps clear lactate even more. And just to clear up a myth: that post-workout soreness (DOMS) isn't from lactic acid—it's from tiny muscle tears and inflammation (*Lactate Shuttle Study*).

SECTION THREE: MICRO-WORKOUTS AND TRACKING

You don't need a gym to stay strong and fit—small bursts of movement and smart tracking can keep you moving like you mean it, every day.

It's mainly about self-discipline, consistency, and building new habits. My friend, former *Major League Baseball* All-Star right fielder for the *L.A. Dodgers* **Yasiel Puig** said:

"You must be more disciplined daily to get better and learn the game."

Tad with Major League Baseball All-Star Yasiel Puig

Source – Sisler Private Collection

4.3.1: STRENGTH SNACKS

Micro-workouts, or "strength snacks," are quick ways to build power without a gym.

- Doing 20 air squats before meals can lower your blood sugar spike by 25%, keeping energy steady.
- Turn grocery runs into a stair-dash for a free HIIT session, or hold an isometric lunge during a phone call to engage 200 muscles at once.
- Try a push-up ladder, adding one rep each day to compound your strength over time.

An accountant added 12% to his grip strength just by sneaking these moves into his day, proving you can get stronger without stepping foot in a gym (*Air Squats Glucose Study*).

4.3.2: NEAT AND DAILY MOVEMENT

Non-Exercise Activity Thermogenesis (NEAT) is the hidden calorie burn from everyday movements, and it adds up fast. Using a standing desk with a rocker board can burn an extra 150 calories per work shift, keeping you active without thinking about it. Walk-and-talk meetings not only burn calories but also spark creativity, making work more fun (I've noticed a few organized crime figures using these walk-and-talk meetings regularly, talking in hushed tones, but for different reasons). Fidgeting—like tapping your foot—can burn up to 350 calories a day, according to Mayo Clinic research (*NEAT and Fidgeting*). Set your phone to vibrate every 45 minutes as a cue to move for 60 seconds, like a quick stretch or pace, to keep NEAT high.

4.3.3: TRACKING AND PROGRESSIVE OVERLOAD

Tracking your movement helps you get stronger over time without overdoing it. If you're lifting weights: use a velocity-based app to aim for a bar speed of 0.7 meters per second during lifts, ensuring you're building power (A bar speed of **0.7 m/s (≈ 2.3 ft/s)** typically corresponds to a **moderate to heavy lift**, around **70–85% of your one-rep max** [1RM] depending on the lift).

If your heart-rate variation (HRV) drops more than 10%, it's time for a deload week to recover and avoid burnout. Every month, check your waist-to-height ratio and a strength score—like how many push-ups you can do—to see progress. An AI dashboard can tweak your macros based on your training load, keeping your diet on point. You can turn that emotional rocket fuel into raw physical power.

CHAPTER FIVE
EAT, DRINK AND FAST FOR 120 DAYS

Could skipping just one evening snack slow your biological clock by a year? Research suggests that how and when you eat can have a huge impact on aging, with studies showing that strategic fasting and nutrient timing can roll back markers like insulin resistance and inflammation. In this chapter, we'll explore a longevity-focused way to eat, smart fasting techniques that act like a reset button for your body, and practical ways to time your meals while tracking what works for you.

SECTION ONE: FOUNDATIONS OF LONGEVITY NUTRITION

What you eat sets the stage for how long and how well you live. Let's dig into the core principles of eating for a century-plus, backed by science and real-world wisdom.

For a deeper dive into this subject, check out my book **The Ultimate AI Diet - Consolidating the Best Diets Over the Last 100 Years: Achieve Fast, Lasting Weight Management with a Personalized, Modern Approach Backed by a Century of Proven Success.**

5.1.1: MEDITERRANEAN-PLUS BLUEPRINT

The Mediterranean diet gets a longevity upgrade with a 60% plant, 20% healthy fats, and 20% protein split, validated by the PREDIMED study for its heart and brain benefits (*PREDIMED Study*).

• Extra virgin olive oil (EVOO) is a star here—its polyphenols switch on SIRT1, a protein that repairs your mitochondria, keeping your cells young.

• Wild-caught fish, rich in omega-3s, can reduce telomere shortening by 30%, helping your DNA stay intact longer.

• Aim for the "rainbow rule"—five different veggie colors a day—to feed a diverse gut microbiome, which fights inflammation.

A 103-year-old Sardinian shepherd credits his long life to "olive oil on everything," a habit that's kept him active and sharp well past 100.

5.1.2: PROTEIN AND mTOR CYCLING

Protein is key for muscle, but too much can speed aging by over-activating mTOR, a growth pathway—hitting a leucine threshold of about 2.7 grams per meal (like a chicken breast) sparks muscle repair without overdoing it. Balance this with a strategy: high-protein days when you train, and a low-protein "Protein Wednesday" to let mTOR rest, slowing aging (*mTOR and Aging*). Plant proteins like beans absorb better with vitamin C—add lemon to your hummus for an iron boost. Collagen, great for joints, should be timed away from leucine-heavy meals to avoid competing signals. Keep your IGF-1 levels (Insulin-like Growth Factor 1 is a hormone your body makes that helps control growth, muscle building, and how your cells repair and age—kind of like your body's "youth signal") in the sweet spot—too low leads to frailty, too high raises cancer risk—so aim for moderation with a balanced protein chart.

5.1.3: ANTI-INFLAMMATORY HYDRATION

Staying hydrated is a longevity hack—aim for about 35 ml of water per kilogram of body weight (a little more than half an ounce per pound of body weight) daily to keep your lymph system flushing out toxins.

• Green tea's catechins can drop inflammation marker CRP by 19%, giving your cells a break (*Green Tea and CRP*).

• Balance your electrolytes with a sodium-to-potassium ratio of 1:2—think a pinch of salt with a banana—to support nerve health and blood pressure.

• A pre-meal "digestive shot" of lemon and apple cider vinegar can flatten your glucose peak by 30%, aiding digestion.

An ultra-runner swore by a homemade hibiscus-salt drink to avoid cramps during races, keeping her hydrated and energized for hours.

"Water is the driving force of all nature." - Leonardo da Vinci

Leonardo da Vinci

Credit – ChatGPT 4o generated image

SECTION TWO: FASTING AS INTELLIGENT STRESS

Fasting shouldn't be about starving—it's a controlled stress that can reset your body's aging clock. Let's look at how to fast smartly for maximum benefits.

5.2.1: INTERMITTENT FASTING MENU

Intermittent fasting (IF) comes in flavors to fit your life, like the 16:8 method—eating in an 8-hour window—which can boost your growth hormone fivefold overnight, helping repair and rebuild (IF Growth Hormone Study).

• The 14:10 window is gentler for women or athletes with pregnancy history, balancing benefits with hormonal health. A

• 5:2 approach—two 500-calorie days a week—can cut insulin levels by 25% in 12 weeks, improving metabolic health.

• You can sip black coffee or tea under 5 calories without breaking your fast, but cream will kick you out of it.

A busy CEO kept his muscle mass using 16:8 fasting paired with resistance bands, proving you can stay strong even with a packed schedule. The 16:8 method isn't tough for me in general. I'll eat my dinner around 6PM and I'm really not hungry until mid-morning the next day, so it's entirely manageable.

5.2.2: FASTING-MIMICKING DIET (FMD)

The Fasting-Mimicking Diet (FMD), developed by **Valter Longo**, involves five days at around 800 calories with 10% protein, tricking your body into fasting mode while still eating.

On day 6, you get a stem-cell rebound effect—seen in both mice and humans—that regenerates tissues (*FMD Stem-Cell Study*). You can DIY it with a grocery list of olives, macadamias, veggie broth, and kale chips for variety. It's not for everyone—skip it if you're pregnant, have type I diabetes, or are frail, and always get your doctor's okay. By day 3, the autophagy gene LC3 doubles, clearing out old cellular junk to keep you running like new.

5.2.3: HORMETIC BEVERAGE BOOSTERS

During a fast, certain drinks can amplify the benefits by acting as hormetic stressors—small challenges that make you stronger.

• Green coffee's polyphenols ramp up AMPK, a longevity pathway that keeps your cells efficient.

• Berberine, at 500 mg, mimics metformin's glucose-stabilizing effects, keeping your blood sugar steady (*Berberine and Glucose*).

• Mineral water with bicarbonate helps maintain your body's pH during longer fasts, and electrolyte tabs without sugar prevent heart rhythm issues.

When you break your fast, start with a handful of walnuts and steamed greens to ease your system back into eating, ensuring a smooth transition.

SECTION THREE: PRACTICAL MEAL TIMING AND TRACKING

Eating for longevity isn't just about what you eat—it's about when and how you track it. Let's set up simple strategies to optimize your meals and see real results.

5.3.1: GCM-GUIDED FUEL PARTITIONING

A continuous glucose monitor (CGM) lets you see how meals affect your blood sugar—post-meal levels above 140 mg/dL for over 30 minutes signal a "spike" that can stress your system. Walking for 10 minutes after eating can drop that spike by 30%, keeping your glucose steady (*Post-Meal Walking Glucose*). Experiment with evening versus morning carbs and graph the results to find what works for you. Devices like *Lumen* or *Biosense* measure breath acetone to show when you're burning fat, giving you a clear picture of your metabolic state. A shift-worker flattened her glucose curve by eating her biggest meal earlier in the day, proving timing can make a big difference.

5.3.2: 72-HOUR JUMP START

Let's kick things off with a three-day plan to get your eating on track.

- Start by using a CGM or finger-stick to test your blood sugar three times a day, logging what you eat to spot patterns.
- Set a 12-hour eating window—like 8 a.m. to 8 p.m.—to give your body a daily mini-fast, which can improve insulin sensitivity right away.
- Swap a processed snack for raw almonds and 85% dark chocolate, a combo that keeps you full and steady.
- Drink 500 ml of water first thing in the morning with a pinch of sea salt to jumpstart hydration, like **Jennifer Aniston** does to stay energized (*Jennifer Aniston's Hydration*).
- After each meal, rate your energy on a 1-to-10 scale to see what fuels you best.

Jennifer Aniston

Credit – Wikimedia Commons

Remember though, like **Jennifer Aniston**: when you're incorporating fasting or a new meal plan or anything else into your life, do it with gusto but don't let it consume you past the point of enjoying your life experience. **Jennifer** said:

"Look I eat really well and I work out, but I also indulge when I want to. I don't starve myself in an extremist way. You're not taking away my coffee or my dairy or my glass of wine because I'd be devastated."

5.3.3: LIFELONG INTEGRATION

Like **Jennifer Aniston** said, to make longevity eating a lifestyle, mix things up over time. Rotate quarterly between intermittent fasting, a fasting-mimicking diet, and a protein-recalibration week to keep your body adaptable. Give yourself "cheat" days within reason here and there. Use seasonal produce to diversify your gut microbiome—check a farmers-market calendar to plan meals with variety. Maybe host a monthly "longevity supper club" potluck with friends to make healthy eating social and fun, boosting your adherence.

Every year, get a full lab check with a lipid panel, fasting insulin, omega-3 index, and DNAmAge to track your progress. Food and fasts fuel you, but nothing rebuilds you like sleep—now let's talk about engineering nights that roll back the clock.

CHAPTER SIX

SLEEP YOUR WAY YOUNGER

Lose 90 minutes of deep sleep tonight and tomorrow your cells act 10 years older—that's how much sleep matters to your body's aging clock. Studies show that poor sleep ramps up inflammation and messes with repair processes, but getting it right can make you feel and function younger. In this chapter, we'll explore how to sync your body's rhythms, build rituals to fall asleep faster, and track your progress so every night helps you wake up fresher and healthier.

SECTION ONE: CIRCADIAN CODING

Your body runs on an internal clock that controls everything from repair to energy levels. Let's get that clock ticking in your favor for better sleep and slower aging.

6.1.1: CLOCK GENES AND REPAIR

Your body has clock genes like PER and BMAL-1 that flip the switch on over 3,000 repair genes every night, fixing DNA and clearing out cellular junk while you snooze. But blue light from screens after sunset can delay melatonin—your sleep hormone—by two hours, throwing off this repair schedule. Eating within a set window, like 10 hours a day, syncs your liver's clock and can lower your risk of fatty liver disease by keeping digestion in rhythm. Getting morning sunlight—at least 1,000 lux, about 10 minutes outside—resets your cortisol decline curve, helping you feel alert all day (*Morning Light and Cortisol*). A night-shift nurse turned her pre-diabetes around by mimicking dawn with a bright lamp each morning, syncing her body clock despite her odd hours.

6.1.2: CHRONOTYPE HACKS

We all have a chronotype—whether you're a morning lark or a night owl—and knowing yours helps you sleep better; you can figure it out with the free **Munich Chronotype Questionnaire** online.

- If you're an owl, cut off caffeine by early afternoon and use a dawn-simulating light box to wake up easier, since your body naturally wants to stay up late. Larks traveling west should shift bedtime 30 minutes earlier each day to adjust smoothly.
- Magnesium-threonate can help owls get more slow-wave sleep, the deep kind that repairs your brain. For jet lag, try a 12-hour fast, then eat breakfast at your new time zone's morning to reset your clock fast (*Chronotype and Sleep*).

6.1.3: SLEEP ARCHITECTURE UPGRADE

Good sleep isn't just about hours—it's about the right mix of stages, aiming for at least 20% REM for dreaming and 25% deep sleep for repair, which you can track with a ring or band like an *Oura* or *Fitbit.*

- Taking 3 grams of glycine before bed lowers your core temperature by 0.7°C, making deep sleep easier to slip into.
- Mouth-taping—if you don't have sleep apnea—can cut micro-arousals in half by keeping you breathing through your nose.
- White noise, like a fan hum, boosts REM density, especially if you live in a noisy city. As a young father who worked nights, there was constantly noise in my house from my four beautiful children. I woke up early with them and at some point, I took a nap before working late into the night. The hum of my fan (on high-speed to cut out all the noise!) was a lifesaver for me.

• Using red light at dusk, instead of blue, can raise melatonin levels by 5%, setting you up for a better night (*Melatonin and Red Light*).

SECTION TWO: RITUALS AND TECH FOR NIGHTLY RENEWAL

Great sleep starts with how you wind down, and you can use simple rituals or high-tech tools to make every night a mini-rejuvenation session. Let's set up your routine for better rest.

6.2.1: THREE-HOUR WIND-DOWN

Start your wind-down three hours before bed to help your body ease into sleep mode. Finish dinner by the three-hour mark to stabilize your blood sugar—no late snacks! At two hours out, dim your lights to under 50 lux and use amber bulbs to avoid blue light that keeps you wired. One hour before bed, do a quick journal brain-dump—write down whatever's on your mind to clear it out and signal your calming parasympathetic system to kick in. With 30 minutes to go, try 4-7-8 breathing: inhale for 4 seconds, hold for 7, exhale for 8, which can drop your heart rate by 10%. Right before lights out, think of something you're grateful for as your mantra to drift off peacefully.

"Your mind will answer most questions if you learn to relax and wait for the answer." – William S. Burroughs

6.2.2: HIGH-TECH SLEEP SANCTUARY

Tech can take your sleep to the next level by creating the perfect environment. A Chili-Pad set to 18°C (64.4°F) can increase deep sleep by 15% by keeping your body cool, which helps you rest deeper (*Chili-Pad Deep Sleep*). Binaural beats at 0.5 Hz strengthen delta waves, the kind that dominate deep sleep, making it easier to get restorative rest. Smart PJs with silver threads reflect far-infrared (FIR) warmth to relax your muscles as you sleep. A bedtime app can auto-kill your Wi-Fi and dim house lights to remove distractions. An athlete boosted his VO_2max by 12% after overhauling his sleep setup with these tools, showing how tech can supercharge recovery.

6.2.3: POWER NAPS AND POLYPHASIC OPTIONS

Naps can be a game-changer if you do them right. NASA found that a 26-minute nap boosts alertness by 34%, giving you a quick recharge without grogginess (NASA Nap Study). A siesta-style split sleep, with a nap in the afternoon, keeps your growth hormone pulsing to rebuild your body. Try the caffeine-nap trick: sip an espresso, nap for 20 minutes, and wake up as the caffeine kicks in for a double boost. Extreme polyphasic schedules like *Uberman*—six 20-minute naps a day—are only good for short projects, not long-term, due to health risks.

After any nap, get some sunlight to reset and shake off any lingering grogginess.

SECTION THREE: TRACKING, TROUBLESHOOTING, AND INTEGRATION

To make sleep your superpower, you'll want to track how well it's working, fix any issues, and build habits that last. Let's make sure every night counts toward a younger you.

6.3.1: BIOMARKERS OF NIGHTLY REPAIR

Tracking your sleep's impact on your body helps you know if you're on the right path. A drop in your morning resting heart rate means you're recovering well—your body's thanking you for the rest. If your **HRV** (**Heart Rate Variability** is a measure of the tiny fluctuations in time between your heartbeats), measured as **RMSSD** (Root Mean Square of Successive Differences, a time-domain measure of heart rate variability that reflects short-term variations between heartbeats, primarily indicating **parasympathetic or rest-and-digest nervous system activity**), dips below 40 milliseconds, it's a sign to take it easy that day, as stress might be catching up.

• Deep sleep also ties to fasting insulin levels—the better you sleep, the better your insulin works.

• Poor sleep for even two nights can spike inflammation marker IL-6, so keep an eye on it with lab tests.

• Slow-wave sleep clears β-amyloid, a brain toxin linked to Alzheimer's, so if you're feeling foggy, your deep sleep might need a boost (*Sleep and Inflammation*).

6.3.2: 72-HOUR JUMP-START

Let's get your sleep on track with a quick three-day plan.

• Set your phone to sunset mode and ditch screens by 9 p.m. to cut blue light—maybe swap your late-night TV for a paperback with a red-light lamp to keep things cozy.

• Add a 10-minute sunwalk each morning to reset your clock, like **Matthew McConaughey** does to start his day fresh (*Matthew McConaughey's Morning Routine*).

• Sip a glycine tea tonight—3 grams in warm water—to help you slip into deep sleep, and log your sleep score tomorrow with a tracker to see the difference.

• For the next three nights, skip alcohol completely, as even one glass can mess with your REM, and watch how much better you feel.

"Without exercise, I don't feel like my head works right."
- Matthew McConaughey

Matthew McConaughey
Credit – Wikimedia Commons

6.3.3: LIFELONG INTEGRATION

To keep your sleep game strong for life, build habits that evolve with your needs. Do a monthly "sleep audit" to check your REM and deep sleep percentages against your goals, tweaking as needed. Adjust for seasons—use red light in winter to mimic dusk and blackout curtains in summer to block early sun. If apnea risk factors like snoring or weight gain creep up, get a quarterly polysomnography test to catch issues early. Share your wind-down ritual with your partner to stick to it together, making it a team effort that boosts consistency. Well-slept cells are ready for genetic upgrades—next we'll discover how to rewrite your epigenetic code.

CHAPTER SEVEN
HORMONE HARMONY
CONDUCTING YOUR INTERNAL SYMPHONY

A teaspoon of hormones can steer the health of a 200-pound body—are yours playing Mozart or heavy-metal noise? These tiny chemical messengers control everything from your energy to your waistline, and when they're out of whack, you feel it—think brain fog, stubborn fat, or low drive. In this chapter, we'll map out your body's hormone orchestra, share everyday tricks to keep them in tune, and explore advanced tools to fine-tune your system for a healthier, longer life.

SECTION ONE:
MAPPING THE ENDOCRINE ORCHESTRA

Your hormones work together like a symphony, and understanding their roles helps you keep the music harmonious. Let's break down the key players and how they shape your health.

7.1.1: HPA & HPT MASTER CONDUCTORS

The **HPA axis** (short for **Hypothalamic-Pituitary-Adrenal axis**) is your body's central stress response system. It's like a communication highway between your brain and adrenal glands, coordinating how you respond to stress—both short-term (like a sudden scare) and long-term (like chronic anxiety or burnout).

Your HPA axis pumps out cortisol to handle emergencies, but when it's stuck on, you end up with belly fat and memory fog from too much stress hormone floating around. The **HPT axis** stands for the **Hypothalamic–Pituitary–Thyroid axis**—it's your body's internal thermostat for **energy, metabolism, and temperature regulation**.

The HPT axis, led by your thyroid, sets your metabolic speed; if T_3 levels dip, you might feel cold and sluggish all the time. A blue-light dawn lamp paired with slow, paced breathing can normalize your morning cortisol curve, waking you up without the jitters. Wearable heart rate variability (HRV) straps, like those used by tech mogul **Elon Musk** to monitor stress, can now flag cortisol spikes in real time (*Cortisol and HRV Study*). A busy executive reversed her adrenal fatigue by using a dawn-sim lamp and breathing exercises each morning, getting her energy back in weeks.

7.1.2: SEX HORMONES ACROSS LIFE STAGES

Sex hormones aren't just for one gender—testosterone helps women keep strong bones and a healthy drive, while estrogen protects men's arteries from plaque buildup. Resistance training, like lifting weights, can boost testosterone by 15% for up to 48 hours in both men and women, giving you a natural lift. After menopause, women's estrogen drop doubles their heart disease risk, but phytoestrogens from foods like soy can help bridge the gap. Missing sleep flattens your LH pulse (An LH pulse is a short burst of luteinizing hormone released by the pituitary gland, crucial for regulating ovulation in women and testosterone production in men), which tanks your sex hormones, so prioritize those Z's.

DHEA, a vitality hormone, supports libido, but always test your levels before supplementing to avoid overdoing it (*Sex Hormones and Training*).

"Sex without love is merely healthy exercise." - Robert Heinlein

7.1.3: APPETITE TRIAD—INSULIN, LEPTIN, GHRELIN

Your appetite hormones—insulin, leptin, and ghrelin—can make or break your eating habits. A high-fructose diet, like too many sugary sodas, can lead to leptin resistance, where your brain thinks you're starving even after a big meal. Ghrelin, your hunger hormone, spikes five hours after eating, but high-intensity interval training (HIIT) can blunt that spike by 27%, keeping cravings in check. A splash of apple cider vinegar before meals can drop insulin spikes by 30%, helping you avoid blood sugar roller coasters. Using a continuous glucose monitor (CGM) reveals which foods send your levels soaring—think sneaky culprits like white bread. A college athlete used CGM cues to stop late-night binges, finding that oatmeal triggered fewer cravings than cereal (*Insulin and Vinegar Study*).

SECTION TWO: LIFESTYLE LEVERS TO RETUNE HORMONES

You don't need a lab to balance your hormones—everyday habits can make a big difference.

Let's look at simple ways to tweak your diet, workouts, and stress to keep your system humming.

7.2.1: NUTRIENT TIMING AND MACRO CYCLING

Timing your food can help your hormones play nice—think of it as giving them the right cues at the right time. Start your day with protein at breakfast, like eggs or Greek yogurt, to spark dopamine and keep you focused, then save carbs for dinner to boost serotonin and help you sleep. A 14:10 fasting window—eating within 10 hours—can increase your growth hormone fivefold, supporting repair while you rest. Once a week, refeed with carbs to keep your thyroid from slowing down, especially if you're cutting calories. A shot of olive oil before meals bumps up GLP-1, a satiety hormone that helps you feel full, and eating cruciferous veggies like broccoli at lunch clears excess estrogens thanks to sulforaphane (*Nutrient Timing Study*).

Suzanne Somers was my friend. We lived in the Palm Springs, California, area and worked many charity events together. **Suzanne** was known for her role on *"Three's Company,"* a popular sitcom. She was an amazing actress and a very kind woman, and I was blessed to know her. Her ideas on food, hormones, and fitness were groundbreaking at the time. **Suzanne** said:

"To successfully lose weight after 40, you need to understand that imbalanced hormones and chemicals have a huge impact on your health. It is possible, however, to live a clean, healthy, and long life - with a slim body, even after 40!"

Suzanne Somers

Credit – Picryl / creativecommons.org

7.2.2: TRAINING, HEAT AND COLD

Your workouts and temperature tricks can give your hormones a natural boost without any fancy supplements. Compound lifts, like squats or deadlifts, spike testosterone for an hour after your session—think of how **Hugh Jackman** transformed for Wolverine with heavy lifting (Hugh Jackman's Training).

Zone-2 cardio, like a brisk walk, raises adiponectin, a hormone that helps burn fat. A 20-minute sauna session at 90°C (194°F) twice a week can increase growth hormone fivefold, giving your body a repair boost. A cold plunge under 12°C (53.6°F) spikes norepinephrine by 200%, firing up brown fat to burn calories. Blood-flow-restriction (BFR) bands, used with light weights, trigger local IGF-1 (Insulin-like Growth Factor 1) to build muscle without heavy loads (*Sauna and Growth Hormone*).

7.2.3: STRESS-TAMING RITUALS

Stress can throw your hormones into chaos, but simple rituals can bring them back in line.

• A 10-minute VR nature scene—like a forest walk—can drop cortisol by 25%, giving your body a break from fight-or-flight mode.

• Ashwagandha, at 600 mg daily, can normalize thyroid hormone TSH in folks with mild hypothyroidism, boosting energy.

• Heart-coherence breathing, where you sync your breath to a calm rhythm, raises DHEA, your vitality hormone, helping you feel more alive.
• Laughter yoga, where you fake giggles until they turn real, lifts oxytocin and calms your nervous system.

My niece **Brynn,** a busy mom, swapped her nightly doom-scrolling for a sunset walk, cutting her evening cortisol in half and feeling more balanced (*Cortisol and VR Nature*).

SECTION THREE: ADVANCED BOOSTERS AND INTEGRATION

For those ready to take hormone health to the next level, let's explore some advanced tools and a plan to keep your system in tune for the long haul.

7.3.1: PEPTIDES AND PRECISION HRT

Advanced tools like peptides and precision hormone replacement therapy (HRT) can fine-tune your system, but they need careful handling. BPC-157, a peptide, speeds up tendon repair by boosting VEGF, a growth factor that heals tissues. Tesofensine curbs appetite without the jitters, but you'll need a doctor's supervision to use it safely. For men, trans-scrotal testosterone cream avoids the estradiol spike you get with injections, keeping things balanced. Women can use rhythmic estrogen-progesterone cycling to mimic a youthful pattern, reducing side effects.
AI-powered dried-blood-spot kits can auto-adjust HRT micro-doses weekly, making it easier to stay in the sweet spot without guesswork (*Peptides and Healing*).

7.3.2: 30-DAY HORMONE RESET PLAN

Let's reset your hormones with a 30-day plan that anyone can follow.
• In week 1, get baseline labs—check your CBC, free T3/T4 for thyroid, fasting insulin, and testosterone-to-estrogen ratio to know where you stand.
• For weeks 2 and 3, cycle your macros with strength training on Monday, Wednesday, and Friday, and Zone-2 cardio on Tuesday and Thursday to balance hormones like adiponectin and testosterone.
• Each day, do 10 minutes of mindfulness, aiming for an HRV above 60 milliseconds to keep stress in check.
• By week 4, repeat your labs to compare trends and see how far you've come.

My friend **Elisa**, who just retired, dropped her biological age by three years in just 30 days using this plan, thanks to better sleep and smarter eating.

7.3.3: TRACKING AND LIFELONG TUNE-UPS

To keep your hormones in harmony for life, tracking and small adjustments are key.

• Each month, use a symptom scorecard to rate your energy, libido, and sleep—think of it like a quick check-in with your body.
• Every three months, do a full lab workup and a week of CGM tracking to tweak your diet and training based on your insulin and glucose trends.
• If your HRV drops more than 10%, take it easy and add some extra carbs to help your body recover.
• Use a wearable to see how your mood ties to hormone-friendly habits, like eating protein at breakfast.

Now that we have a plan for hormone harmony, let's nurture the messenger hub in your belly—your microbiome.

CHAPTER EIGHT
YOUR GUT-IMMUNE ALLIANCE

Did you know that just two pounds of microbes outnumber all of your human cells? Are these microbes your friend or saboteur? That's right, your gut is home to trillions of tiny organisms that can either keep you healthy or stir up trouble, depending on how you treat them. They influence everything from your mood to your immune system, and in this chapter, we'll get to know these little helpers, learn how to make them your allies for a longer life, and set up a plan to keep them happy while tracking the results.

SECTION ONE:
MEET YOUR MICROBIAL COMMAND CENTER

Your gut is like a bustling command center, packed with microbes that talk to your brain and immune system.

For a greater look into this subject and supplements that can assist you, check out my book **Vitamins, Supplements, and Herbs for Health and Longevity: Boost Your Immunity, Increase Energy, and Feel Younger in Minutes a Day.** Let's break down who's in there and how they're running the show.

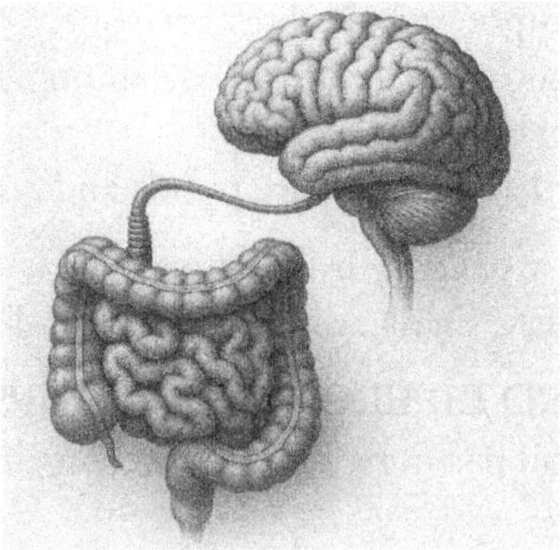

8.1.1: MICROBIOME BASICS

Your gut houses about 100 trillion organisms with 3 million genes—that's 150 times more than your human genes! The balance between two main groups, Firmicutes and Bacteroidetes, can predict your obesity risk; too many Firmicutes might make weight harder to manage.

Good bugs that produce butyrate help tighten your gut lining and fuel your colon cells, keeping things healthy down there. But when the balance goes off—called dysbiosis—it lets a toxin called LPS sneak into your bloodstream, sparking inflammation all over your body (*Microbiome and Inflammation*).

ALEXANDRA BURKE

British singer **Alexandra Burke** (winner of *The X Factor*) spent years performing through stage lights and stomach knots. Offstage she curled up with bloating so bad "I looked six months pregnant," she told reporters.

Working with a nutritionist, she began a daily routine of kefir, plant fiber, and stress-taming breathing drills that lower gut-brain tension. The result: fewer flare-ups, stronger vocals, and the confidence to headline West End shows—all because she learned to "feed the good bacteria first."

Alexandra Burke
Credit – Wikimedia Commons

8.1.2: GUT-BRAIN-IMMUNE HIGHWAY

Your gut, brain, and immune system are chatting all the time through a highway called the vagus nerve, which sends microbe messages straight to your brain's control center, the hypothalamus. About 90% of your serotonin—the feel-good chemical—is made in your gut, affecting your mood and even how well your digestion moves along.

Short-chain fatty acids (SCFAs) from gut bugs help calm your immune system by supporting T-reg cells, which keep autoimmune reactions in check. If your gut lining gets leaky—thanks to a protein called zonulin—allergens can slip through, setting off a storm of immune reactions. Stress can mess up your gut microbes in just 24 hours, with cortisol favoring the bad guys, so staying calm will help your digestive system (*Gut-Brain Axis*).

8.1.3: HIDDEN AGITATORS

Some everyday things can quietly mess with your gut's balance, and you might not even realize it. Emulsifiers like polysorbate-80, found in processed foods, can thin your gut's protective mucus layer—at least in mice studies. Glyphosate, a common pesticide, disrupts a pathway that good bugs like lactobacilli need, leaving you with fewer of the helpful ones.

Too much alcohol can wipe out Akkermansia, a microbe that keeps your metabolism on track, while chlorinated tap water might lower your gut diversity compared to filtered water, according to a 2023 study (*Chlorinated Water and Gut Diversity*). And antibiotics? Just one course can shrink your gut diversity by 25% for up to six months, so use them wisely.

SECTION TWO: CULTIVATING A LONGEVITY-FRIENDLY GUT

Now that you know who's in your gut, let's grow the good guys and keep the troublemakers in check with some simple, everyday habits.

8.2.1: PRE- AND PROBIOTIC GAME PLAN

Feeding your gut the right stuff is like planting a garden that grows health. Prebiotics like inulin and resistant starch—think onions or cooled potatoes—nourish butyrate-producing bugs that keep your gut lining strong. A "synbiotic stack" pairing probiotics like L. reuteri with galacto-oligosaccharides (GOS) gives your microbes a double boost.

Fermented foods such as kimchi and kefir add live cultures and polyamines, which help your cells stay young. Spore-based probiotics like Bacillus are tough enough to survive your stomach acid, making them perfect for travel.

A world-touring entrepreneur stayed symptom-free on the road by packing spore-based probiotics, keeping her gut happy no matter where she was (*Probiotics and Gut Health*). Although I'm not officially endorsing the following company here, I personally use a Prebiotic/Probiotic from *Seed Health,* which works very well for me.

8.2.2: POLYPHENOLS AND POST-BIOTICS

Polyphenols are like superfoods for your gut microbes, and they're found in some of your favorite treats.

• Pomegranate gets turned into urolithin A by your gut bugs, which revs up mitophagy—the process that cleans out old mitochondria to keep your cells running smoothly.

• Green tea's catechins help balance your oral bacteria, so you swallow less of the bad stuff that can cause trouble.

• Dark chocolate's flavanols feed Roseburia, a microbe that improves insulin sensitivity, keeping your blood sugar steady (*Polyphenols and Insulin Sensitivity*).

• Post-biotics like butyrate are being studied for colitis relief through enemas, giving your gut a direct dose of healing.

• You can even use an AI grocery scanner app to check the "microbe score" of foods by scanning barcodes, making shopping a breeze.

8.2.3: LIFESTYLE LEVERS

Your daily habits can shape your gut just as much as what you eat. A 10,000-step walk in a forest boosts your gut's alpha-diversity—think of it as adding more variety to your microbial garden—compared to an indoor treadmill session.

• Sticking to a 12-hour eating window, like noon to midnight, syncs your microbes' circadian rhythm, keeping them happy and balanced.

• Stress-relief breathwork, like slow belly breathing, ramps up salivary IgA, an immune defender in your gut.

• Getting your hands dirty with pets or gardening introduces you to soil microbes, often called "old friends," that help your immune system stay strong.

So, pet your dog and plant some flowers! I promise that will help your optimistic outlook too! Regular saunas can flush out endotoxins, while a cold plunge after cools inflammation, giving your gut a double win (*Nature Exposure and Gut Diversity*).

8.2.4: THE GUT-BRAIN AXIS: YOUR SECOND BRAIN'S LONGEVITY LINK

In Section 8.1.2, we talked about how your gut isn't just for digesting food—it's talking to your brain all the time through something called the gut-brain axis, like a secret hotline between your belly and your head using your vagus nerve and hormones to send messages both ways—so a happy gut can make you feel calm and sharp, while a messed-up gut can leave you anxious or foggy, speeding up aging with inflammation. We've talked about feeding your gut good stuff like fiber and fermented foods (Sections 8.2.2 and 8.2.1), but here's the cool part: as your balanced gut microbiome lowers stress hormones like cortisol, it also boosts feel-good chemicals like serotonin—yep, 90% of your serotonin is made in your gut! That means your gut health directly affects your mood, which we know impacts how long you live (Chapter 14).

Studies—like one from Harvard in 2023—show that people with diverse gut bacteria have better mental health and lower inflammation, which helps fight off age-related diseases. So, when you're eating that kale salad or sipping on kefir, you're not just feeding your body—you're feeding your brain, too, helping it stay resilient and keeping your whole system young. This gut-brain link is why we've been so focused on your "second brain" in this chapter—it's a key player in living longer, naturally, by keeping your mind and body in sync.

"Calm mind brings inner strength and self-confidence, so that's very important for good health." – Dalai Lama

SECTION THREE: TESTING, TRACKING AND REJUVENATION

Let's make sure your gut stays in top shape by testing what's going on, tracking your progress, and following a simple plan to keep it thriving for the long haul.

8.3.1: DIY MICROBIOME DASHBOARD

Testing your gut microbiome is like getting a report card for your inner garden, and you can do it at home with shotgun-sequencing kits that map out everything down to the pathways your microbes use. Test every three months to spot seasonal shifts—like how summer fruits might change your bugs—and log your diet to see what's driving those changes.

Overlay your CGM data to connect the dots between what you eat, how your microbes react, and how your blood sugar behaves. Aim for a Shannon Index above 60, which is the diversity sweet spot for longevity (*Microbiome Testing and Diversity*). Keep your raw data safe by encrypting it, and only share summaries with apps to protect your privacy.

8.3.2: PRECISION SYNBIOTICS AND ENGINEERED BUGS

Advanced gut tools can take your health to the next level, targeting specific issues with precision.

- A capsule called SER-109 has been shown to cut recurrent C. diff infections by 88%, restoring balance after antibiotics (SER-109 C. diff Study).

• Scientists are using CRISPR to edit E. coli to deliver IL-10, an anti-inflammatory for colitis, though it's still in mouse studies.

• Pasteurized Akkermansia, a key gut microbe, lowered visceral fat in a 2022 human trial, helping with weight management.
• Phage cocktails can zap harmful bacteria without hurting the good ones, keeping your gut ecosystem intact.
• A pre-diabetic reversed her HbA1c levels using a tailored synbiotic and phage combo, showing how personalized gut care can change your health.

8.3.3: 3-MONTH GUT REJUVENATION PLAN

Let's rebuild your gut with a three-month plan that's easy to follow and packed with benefits.
• Month 1 is your clean slate: ditch ultra-processed foods—like chips and sodas—and aim to eat 30 different plant species a week, think of it as a rainbow challenge with colorful veggies and fruits.
• In month 2, rebuild with a synbiotic stack of pre- and probiotics, plus a 14:10 fasting window—eat from 8 a.m. to 10 p.m.—to give your gut a rest.
• By month 3, optimize with a polyphenol-rich micro-diet, like adding green tea and dark chocolate, and hit the sauna three times a week to flush out toxins—just like **Chris Hemsworth** does to stay in superhero shape (*Chris Hemsworth's Wellness Routine*).

Each week, log your stool form, HRV, mood, and energy to see how your gut's feeling. With your microbial metropolis thriving, it's time to super-charge the command center upstairs—your adaptable, ageless brain.

For certain roles, **Chris Hemsworth** had to gain weight fast, which can do a number on your system. This is why he pays close attention to his gut health. **Chris** said:

"Eating when you're not hungry and taking in that amount of food is exhausting."

Chris Hemsworth
Credit – Flickr / creativecommons.org

CHAPTER NINE
BRAIN GAINS FOR LIFE

Adults who practiced just 20 minutes of targeted brain-training daily grew their memory center by 2% in eight weeks—the same region that normally shrinks with age. That's right, your brain can keep growing and getting sharper, no matter how old you are, if you give it the right kind of workout. In this chapter, we'll show you how to keep your brain flexible, protect it from aging, and build a plan to boost your memory, focus, and smarts for life. Playing *Words With Friends* or doing crosswords are great starts, but there's much more you can do to keep your brain stimulated and high performing.

SECTION ONE: NEURO-PLASTICITY NEVER RETIRES

Your brain isn't stuck—it's like Play-Doh, ready to be reshaped at any age with the right habits. Let's explore how to keep it growing and sharp.

9.1.1: SYNAPSES: USE 'EM, GROW 'EM

Your brain's connections, called synapses, can grow new branches—called dendritic spines—in just hours when you try something new, like learning a dance move or a new recipe.

Aerobic exercise, like a brisk walk, releases a protein called BDNF, which acts like "Miracle-Gro" for your neurons, helping them grow stronger.

Taking 1 gram of DHA—a type of omega-3 found in fish oil—every day can make your synapses snappier, speeding up your reaction time by 9%.

Intermittent fasting gives your brain a spring-clean by boosting a process called neuronal autophagy, clearing out old junk to keep things running smoothly (*BDNF and Exercise*). My friend **Danielle,** a 68-year-old pianist, started learning Spanish and saw her brain's cortex thicken on an MRI, proving you're never too old to grow your mind.

9.1.2: COGNITIVE CROSS-TRAINING

Think of your brain like a gym—mixing up workouts makes it stronger all over. Pairing activities like juggling with learning a new language can thicken both sides of your brain, giving you a mental boost.

Playing music after 50, like **Sting** picking up the lute, can grow the bridge between your brain's two halves—called the corpus callosum—so they talk to each other better (*Music and Brain Connectivity*). I have an excellent MUSIC MASTERY series of books on Piano, Voice, Guitar, Bass, and Drums available on Amazon. These books are great whether you've never touched an instrument or if you're proficient. I highly recommend learning to play an instrument for longevity's sake! You can access these books on Amazon using this QR code:

Virtual reality "exergames," where you move and solve puzzles, spike norepinephrine, a chemical that sharpens your focus and working memory.

Using the Pomodoro technique—25 minutes of work, 5-minute breaks—keeps your brain's thinking center from getting tired. An AI coach can even use EEG data to switch up your tasks when your attention starts to dip, keeping you in the zone.

9.1.3: NOOTROPIC NUTRITION

Food and supplements can give your brain a little extra juice—think of them as brain snacks.

- Creatine, at 5 grams a day, helps vegans boost short-term memory by filling in a nutrient gap.
- Lion's Mane mushroom stimulates a growth factor called NGF, with early studies showing a 15% bump in recall scores (*Lion's Mane and Memory*).
- Blueberries, packed with polyphenols, increase blood flow to your brain, lighting up thinking areas on fMRI scans.
- Pairing caffeine with L-theanine—like in green tea—keeps you alert without the jitters, giving you calm focus.
- Choline from egg yolks feeds acetylcholine, a brain chemical that helps you learn faster, so don't skip breakfast.

SECTION TWO: SHIELDING THE BRAIN FROM DECLINE

Aging doesn't have to mean losing your mental edge—let's look at ways to protect your brain from decline and keep it sharp for years to come.

9.2.1: METABOLIC AND VASCULAR GUARDIANS

Watching my beautiful sister deteriorate from Parkinson's Disease has been awful for my family. Inside, she will always be that vibrant, caring sister, mother, and friend, but her body has been devastated by the disease. Neuro-diseases like Parkinson's and Alzheimer's don't have cures yet, but you can do preventative measures to lower your chances of neurological brain disorders.

When your blood sugar gets out of whack, it starves your brain cells—insulin resistance doubles your Alzheimer's risk if you have type 2 diabetes.

- The Mediterranean-MIND diet, full of veggies, fish, and nuts, slowed cognitive decline by 53% in a big study, keeping your brain healthier for longer (*MIND Diet Study*).
- Zone-2 cardio—like a steady bike ride where you can still chat—floods your brain with oxygenated blood, kind of like giving it a refreshing shower.
- A shot of beet juice, loaded with nitrates, widens blood vessels in your brain for six hours, boosting flow.
- Saunas four times a week can cut your dementia risk by 66%, according to a Finnish study, so don't shy away from a good sweat (*Sauna and Dementia Risk*).

9.2.2: DEFEATING NEURO-INFLAMMATION

Inflammation in your brain can speed up aging, but you can fight it with the right habits.

- Curcumin, a compound in turmeric, blocks a pathway called NF-κB to cool down inflammation, especially in forms that can cross into your brain.
- Getting enough sleep—aim for at least 7 hours—lets your brain flush out β-amyloid, a toxin linked to Alzheimer's, through a cleaning system that kicks in during deep sleep.
- Eating resistant starch, like cooled potatoes, feeds gut bugs that make butyrate, which calms down overactive brain immune cells called microglia.
- Low-dose naltrexone can lower inflammation marker TNF-α, with early trials showing promise for MS.

An ex-NFL player slashed his migraines by following an anti-inflammatory diet and hitting the sauna regularly, proving these habits can make a real difference (*Curcumin and Brain Inflammation*).

"Inflammation is the cornerstone of Alzheimer's disease and Parkinson's, multiple sclerosis - all of the neurodegenerative diseases are really predicated on inflammation." – David Perlmutter

KRISTIN CHENOWETH

Tony-winning singer-actress **Kristin Chenoweth** has battled chronic, ice-pick migraines since her teen years. In a 2019 Healthline interview she explained that ditching trigger foods (sugary sodas and deep-fried snacks) and shifting to what her doctor called an "anti-inflammatory rainbow plate" cut her attacks in half. She also swears by 15-minute sessions in an infrared sauna before bed to "melt the tension" that sparks her headaches. The combo helps her hit high notes eight shows a week without that familiar migraine fog. Kristin is also a big advocate of hydration.

"As much as you can eat healthy, it's also important to remember to drink healthy too. Tea is very healing." – Kristin Chenoweth

Kristin Chenoweth

Credit – Flickr / creativecommons.org

9.2.3: DIGITAL HYGIENE

Your digital habits can either help or hurt your brain, especially when it comes to sleep and focus.

- Wearing blue-light blockers after 8 p.m. protects melatonin, the sleep hormone that helps your brain solidify memories overnight.
- Follow the 20-20-20 rule: every 20 minutes, look at something 20 feet away for 20 seconds to give your eyes—and brain—a break from screen strain.
- Doom-scrolling on your phone can drain your dopamine, making it harder to focus, so set "scroll windows" to limit mindless browsing.
- Keep your phone at least 12 inches away while you sleep to cut down on EMF exposure, which might mess with your brain waves.
- A screen-time app can nudge you with a "cognitive fuel gauge," helping you know when to take a break (*Blue Light and Sleep*).

SECTION THREE: BUILD YOUR COGNITIVE-EDGE PLAN

Let's put it all together with a plan to boost your brainpower, using tools to track your progress and habits to keep your mind sharp for life.

9.3.1: 90-DAY SYNAPSE SPRINT

This 90-day plan is like a gym bootcamp for your brain, helping it grow stronger step by step.

• For the first four weeks, aim for 150 minutes of cardio—like walking or biking—and play a dual n-back game for 15 minutes a day to challenge your memory, just like tech guru **Elon Musk** trains his focus with brain games (*Elon Musk's Brain Training*).

• Weeks 5 to 8, add a language app like *Duolingo* and practice an instrument—like the guitar—three times a week to cross-train your brain.

• In weeks 9 to 12, mix in HIIT workouts, sauna sessions, and a 14:10 fasting window to support brain health with exercise and nutrition.

Every two weeks, test yourself with *Cambridge Brain Sciences* to track your progress. An entrepreneur jumped from the 60th to the 82nd percentile in processing speed by following this plan, showing how small steps can lead to big brain gains.

"I think it's very important to have a feedback loop, where you're constantly thinking about what you've done and how you could be doing it better. I think that's the single best piece of advice: constantly think about how you could be doing things better and questioning yourself."
– Elon Musk

Elon Musk
Credit – Wikimedia Commons

9.3.2: TECH AND WEARABLES

Tech can give your brain a cutting-edge boost, helping you work smarter, not harder.

• An HRV ring can spot your "brain-best" days—when your heart rate variability is high—so you can schedule tough tasks like learning or problem-solving for those times.
• A Muse S EEG headband trains you to boost alpha waves, helping you focus better during work or study.
• A tDCS headset, used safely with a proper protocol, can pair with learning sessions to enhance memory—like how gamers use it to level up their skills.

Future tech like *Neuralink* and *Synchron* is working on brain implants to boost cognition, though they're still in early stages. Quantum MRI is coming down the road, promising to map your brain's tiniest connections for a deeper look at how it's wired (*tDCS and Learning*).

9.3.3: LIFELONG INTEGRATION

To keep your brain sharp for life, make learning and protection a regular part of your routine.
• Every month, do a "skill swap"—rotate hobbies like painting one month and chess the next to keep things fresh and spark new brain connections.
• Every three months, reset with an anti-inflammatory diet—load up on veggies and fish—to keep brain inflammation low, much like actress **Jennifer Garner** does to stay focused and healthy (*Jennifer Garner's Healthy Habits*). Like everyone, Jennifer is always concerned about getting older and how that will impact her health. **Jennifer** said:

"I do think about aging. I have those moments of panic and vanity, but life keeps getting better, so you can't worry about it too much."

Jennifer Garner
Credit – Wikimedia Commons

• Once a year, get a cognitive check-up with a neuropsych battery to track your baseline memory and thinking skills.
• Teach a brain skill—like how to play a song on the piano—to a friend or family member; teaching helps lock in what you've learned.

Your upgraded brain still lives inside an environment—next, we'll let nature become your personal longevity clinic.

CHAPTER TEN
NATURES PRESCRIPTION: REWILDING YOUR BIOLOGY

Just two hours a week in green space cuts all-cause mortality by 12%, yet the average American spends 93% of their life indoors—crazy, right? Nature isn't just a pretty backdrop; it's a powerful way to boost your health, mood, and even how long you live. In this chapter, we'll get you back in tune with the outdoors, tweak your home to bring the wild inside, sync up with the rhythms of seasons and the moon, and even learn from nature's teamwork to make your life better.

SECTION ONE: REUNITING WITH PLANETARY RHYTHMS

Two of my older sisters were "hippies" for a time. Much of the world looked at them like they were freaks, but they were loving people and very much in tune with nature and their inner selves during that period.

Nature actually does have rhythms that your body craves—like sunlight and the earth's energy—and getting back in sync can make you feel amazing. Let's explore how to tap into these natural beats.

10.1.1: SUNLIGHT-HORMONE AND CLOCK REGULATOR

Sunlight does way more than give you a tan—it's like a reset button for your body's clock.
• Just 10 minutes of morning sun at 1,000 lux—about the brightness of a cloudy day—can reset your brain's master clock, helping you fall asleep 37 minutes faster at night.
• Midday sun gives you a dose of UV-B rays, which your skin turns into vitamin D_3, lifting your mood and immune system.
• At sunset, the red and near-infrared light feeds your mitochondria—the energy factories in your cells—making them work better. A quick 15-minute sun break can also boost serotonin, the feel-good chemical, way more than indoor LED lights (*Sunlight and Sleep Latency*).

A software engineer beat seasonal depression by taking "Sun-Salutation breaks" outside each morning, feeling brighter and more energized in weeks.

10.1.2: EARTHING AND SCHUMANN RESONANCE

• Walking barefoot on the ground—called earthing—connects you to the earth's energy, balancing your body's voltage and even thinning your blood a bit to help it flow better.

• Sitting on a grounding mat for an hour can boost your heart rate variability (HRV) by 10%, showing your body's more relaxed. The earth hums at a frequency called Schumann Resonance—7.83 Hz—and it can calm your brain by syncing it to a chill alpha wave state.

• Spending two nights camping can realign your cortisol levels, helping you feel less stressed (this one is tough for me because my idea of camping is staying at the Hilton). Stick to clean grass or sand for grounding, and skip lawns treated with pesticides to keep it safe (*Earthing and HRV*).

10.1.3: FOREST BATHING AND PHYTONCIDES

• Forest bathing—or shinrin-yoku—is just a fancy way of saying "take a walk in the woods," and it can lower your stress hormone cortisol by 15%.

• Breathing in pine scents, like α-pinene, can boost your immune system's natural killer (NK) cells by 40%, and the effect lasts a whole week after your walk. Soil microbes like Mycobacterium vaccae can lift your mood by nudging your brain to make more serotonin, the happy chemical.

• Walking near a waterfall gives you a dose of negative ions, which help clear out your airways (walking barefoot on the beach also produces this same ion effect, along with earthing).

To get the best benefits, use a trail-finder app to pick biodiverse forest paths with a mix of plants for richer air (*Forest Bathing and Cortisol*).

SECTION TWO: DESIGNING BIO-ALIGNED ENVIRONMENTS

You don't have to live in a forest to feel nature's benefits—let's bring the outdoors into your home and daily life with some easy tweaks.

10.2.1: HEALTHY HOME BLUEPRINT

Your home can mimic nature's rhythms to keep you feeling good.

• Use daylight LED bulbs in the morning to wake you up, then switch to amber bulbs at night to wind down without messing with your sleep.

• A HEPA and carbon filter can cut down on tiny air pollutants called PM2.5, lowering your risk of brain aging from dirty air.

• Plants like snake plants and peace lilies clean out nasty chemicals (VOCs) and add a bit of humidity, making your air fresher. In the kitchen, circadian lighting—brighter in the day, softer at night—can stop late-night snacking by keeping your body clock on track.
• Turn off Wi-Fi an hour before bed and hard-wire your workstations to cut down on EMF exposure while you rest (*Indoor Air Quality and Brain Health*).

Some people purchase fresh-air machines; others like negative ion generators. Don't forget that loving dogs and cats activate your feel-good hormones too! My friend, legendary actor **Robert Wagner** said:

"A dog will teach you unconditional love. If you can have that in your life, things won't be too bad."

Robert Wagner and Tad Sisler
Source: Tad Sisler's personal collection

10.2.2: MINIMALISM AND STRESS UNLOADING

A cluttered home can stress you out, raising your cortisol levels, so try a "10-item declutter" ritual—pick 10 things to toss or donate each week.
• Take a break with a digital-fast Sunday, putting away screens to reset your dopamine and feel more refreshed.
• A capsule wardrobe—just a few outfits you love—cuts down on decision fatigue, so you're not stressing over what to wear. My mother used to say, ***"If it doesn't make you feel great wearing it, get rid of it."***

Simplifying your finances, like setting a budget, lowers your overall stress load, leaving you calmer. A family who binned 50% of their stuff saw their sleep and HRV soar, showing how less stuff can mean more peace (*Clutter and Cortisol*).

10.2.3: URBAN NATURE HACKS

Even if you're in the city, you can bring nature closer with some clever hacks.
• Grow a micro-greens garden on your windowsill—it's an easy way to get fresh nutrients and a dose of soil microbes without a backyard.

• Step onto your rooftop or open patio for sun breaks to soak in 5,000 lux of natural light, giving your body a quick reset.
• If you can't get outside, pop on headphones with ocean or forest soundscapes to calm your nervous system.
• VR nature immersion, like a virtual forest walk, can help if parks are hard to reach, boosting your relaxation.
• Volunteering for a trash pick-up event or tree-planting event gives you a social boost and connects you with nature—it's a win-win (*Urban Nature and Stress*).
Get out there and take deep breaths. Live your life with gusto!

SECTION THREE: SEASONAL AND LUNAR SYNCING

Nature's cycles—like the seasons and the moon—can guide your habits to keep your body in harmony. Let's sync up with these rhythms for better health.

Tweaking your habits in new ways is great for those of us who have trouble doing the normal stuff like meditating, practicing mindfulness, or doing yoga. It reminds me of what my friend, *multi-platinum* recording artist **Rod Stewart** jokingly said:

"Surely if God had meant us to do yoga, he would have put our heads behind our knees."

Tad Sisler with Rod Stewart
Source – Sisler Private Collection

10.3.1: CHRONO-BIOLOGY OF THE YEAR

Your body changes with the seasons, so tweak your habits to match.
• In winter, use red-light therapy and omega-3s—like fish oil—to fight the gloom, and go to bed earlier to match shorter days.
• Spring is perfect for a gentle detox with bitters, like sipping dandelion tea, to wake up your digestion.

• Summer's longer days mean you can handle more carbs, especially if you're active, so enjoy that extra fruit.
• In autumn, eat grounding root veggies like sweet potatoes and take more rest to prep for winter. There's even an app that tracks daylight length and adjusts your macros—like carbs and fats—to match the season (*Seasonal Rhythms and Health*).

10.3.2: LUNAR INFLUENCE AND TIDAL HORMONES

The moon's phases can nudge your body in subtle ways, and you can use them to your advantage. At the full moon, melatonin dips a bit, so plan tough workouts for the next morning when you're naturally more alert. During the new moon, your digestion slows down, making it a great day to fast and give your gut a break.

A study of fishermen showed testosterone peaks during the waxing moon, when the moon's getting brighter, so you might feel a bit more energized then.

Biodynamic harvests—crops grown with lunar cycles in mind—have higher nutrient levels, so look for them at farmers' markets.

A triathlete synced his training blocks to lunar phases, cutting his injury rate by timing hard sessions with his body's natural highs (*Lunar Cycles and Hormones*).

10.3.3: 30-DAY NATURE RECONNECTION PROTOCOL

Let's get you back in touch with nature with a simple 30-day plan.
• Each day, spend at least 10 minutes in dawn light—outside if you can—to wake up your body clock, and go barefoot on grass or dirt for an extra grounding boost (watch for sticker-burrs, sharp objects, or broken glass).
• Once a week, plan a 3-hour trip to a green space, like a park or forest, and log how your mood shifts after—maybe you'll feel calmer or happier.
• After lunch, take 15 minutes to ground yourself by standing barefoot outside, letting the earth's energy relax you.
• Create a "biophilic corner" at home with a few plants and a small water fountain for soothing sounds, bringing a bit of nature indoors.

I have a lovely little fountain outside my front window, and I keep it on a timer, so it shuts down at night, reducing power use.

During the day, the calming sound of the fountain is great.

SECTION FOUR: COLLECTIVE WISDOM FROM NATURE

If insects gain strength from collective rhythm, could human communities do the same?

Nature's creatures work together in amazing ways, and we can learn from them to boost our own health and connection. Let's see how teamwork in the wild can inspire us.

10.4.1: BIOMIMICRY AND SWARM WISDOM

I've always thought of humans as the superior beings we are, but I am amazed at how deeply dogs can smell, far more than we ever will comprehend. Animals and bugs across the spectrum each have their own unique magic.

For instance, why is a single ant "dumb," but a colony routinely solves problems IBM super-computers once struggled with? Ants use simple rules—like following scent trails—to make smart group decisions, such as finding the shortest path to food, showing how many small signals add up to one big, clever choice.

Bees do something similar with their "waggle dances," a kind of energy-sharing dance that tells the hive where to find nectar—it's like their version of a high-energy group chat, with ATP as their nectar and brain waves as their dance. We can mimic this by working together in small groups, pooling our ideas to solve problems faster, just like ants and bees do every day (*Ant Colony Decision-Making*).

10.4.2: STARLING MURMURATIONS AND HEART-RATE ENTRAINMENT

Ever seen a flock of starlings swoop and swirl in the sky like a giant, dancing cloud? Those murmurations aren't just beautiful—they show how thousands of birds sync up their movements in real time, a bit like how our heart rates can sync at big events.

At concerts or sporting events, like when you're cheering at a football game, your heart rate can start to match the crowd's energy, creating a shared rhythm that makes you feel connected and alive. It's the same kind of magic that happens when you dance with friends or sing along at a live show—your body tunes into the group's vibe, boosting your mood and even your health (*Heart-Rate Entrainment*).

In fact, I'm surprised more people don't experience heart issues at stadiums during the playoffs!

10.4.3: COLLECTIVE CONSCIOUSNESS IN THE HUMAN EXPERIENCE

Humans can tap into a kind of collective consciousness too, where our shared emotions and intentions create a powerful group effect.

The *Global Love Project* in 2015 brought thousands together to meditate on love and peace, and researchers found a measurable drop in local crime rates during the event (*Global Love Project*).

During the *2017 Women's March,* millions united in solidarity, and studies showed a temporary dip in stress-related health visits in participating cities (*Women's March Impact*).

The 1993 Washington, D.C., meditation experiment saw 4,000 people meditate together, reducing violent crime by 23% over two months, as tracked by local police data (*Washington Meditation Experiment*).

In 2020, a global prayer initiative during the pandemic, involving millions across faiths, correlated with a brief uptick in reported well-being on social media sentiment analysis (*Global Prayer 2020*).

These examples show how coming together with a shared purpose can create a ripple effect, making the world a little better for everyone. Check out my book on how a positive outlook can vastly increase your healthspan and lifespan, entitled **The Science of Positive Thinking: How Mindset, Daily Habits, and Emotional Well-being Can Add Years to Your Life.**

Now that earth and sky power your biology, get ready to fine-tune your cellular engines—next stop: metabolic mastery & mitochondrial power.

CHAPTER ELEVEN
METABOLIC MASTERY AND MITOCHONDRIAL POWER

A single CGM (continuous glucose monitor) study showed that trimming the last 90 minutes of evening eating cut glucose spikes in half within a week—pretty amazing, right? That means small changes to how your body handles food can make a big difference in your energy, health, and how fast you age. In this chapter, we'll break down how to manage your body's fuel like a pro, power up your mitochondria—those tiny energy factories in your cells—and set up an easy plan to keep your metabolism humming for years to come. For a deeper dive into mitochondria and metabolism, check out my book **Stay Healthy, Stay Youthful: The Science of Living to 150: Cutting-Edge Longevity Science for Reversing Aging and Living a Healthier, Longer Life.**

SECTION ONE: FUEL-PARTITIONING 101

Think of fuel-partitioning as deciding when and how your body uses carbs, proteins, and fats—it's like being the boss of your own energy system. Let's dig into some simple ways to make it work better for you.

11.1.1: MACRO-TIMING AND CGM FEEDBACK

Timing your meals can help your body handle food better, and a continuous glucose monitor (CGM) is like a little coach showing you what's going on with your blood sugar. Eat your biggest carb load—like oatmeal or toast—at breakfast when your insulin is most ready to keep sugar spikes in check. If your CGM app shows your blood sugar staying above 140 mg/dL for more than 30 minutes after a meal, that's a "spike" you want to avoid. A quick 10-minute walk after eating can drop that spike by about 30%, keeping things steady. If you notice a morning sugar rise—called the dawn phenomenon—doing some push-ups before breakfast can help tame it (*Post-Meal Walking and Glucose*). A night-shift chef flattened her glucose curve by eating her biggest meal earlier in her shift, showing how re-ordering meals can make a big difference.

My dear friend, legendary actress **Mary Tyler Moore**, struggled with diabetes throughout her life. She had to watch everything she ate, especially sugar. **Mary** said:

> ***"Diabetes is an all-too-personal time bomb which can go off today, tomorrow, next year, or 10 years from now - a time bomb affecting millions like me and the children here today."***

Tad Sisler with Mary Tyler Moore

Source- Sisler Private Collection

The main thing is to do whatever you can to avoid becoming hypoglycemic or diabetic through a proper diet and exercise, if you can.

11.1.2: PROTEIN-LEVERAGE PRINCIPLE

Protein isn't just for muscles—it can help you feel full and keep your appetite in check.

• Aim for about 1.6 grams of protein per kilogram of your body weight each day (about 0.76 grams per pound of body weight)—like a chicken breast or tofu serving—to naturally stop you from overeating.

• Getting 2.7 grams of leucine, an amino acid in foods like eggs, kicks off a muscle-building signal called mTOR, helping you stay strong.

• If you're eating plant proteins, like beans, add a squeeze of lemon for vitamin C to help your body use the iron and lysine better.

• Toss in 10 grams of collagen for your joints, but take it separately from leucine-heavy meals so they don't compete.

• Once a week, try a low-protein "reset day" to keep IGF-1—a growth hormone—in balance without losing muscle (*Protein Leverage Hypothesis*).

11.1.3: FAT-ADAPTATION WITHOUT KETO DOGMA

You can train your body to burn fat for fuel without going full keto, and it's easier than you think. A breath-acetone monitor—like a little gadget you blow into—can show when you're in fat-burning mode, aiming for over 10 ppm. Start with a two-day low-carb primer, cutting back on bread and pasta, to rev up your fat-burning enzymes. Before a high-intensity workout, eat some carbs—like a banana—to keep your thyroid and hormones happy. A spoonful of MCT oil gives you quick ketones for mental clarity on busy days, without needing a strict diet. But watch out—staying very low-carb all the time can raise your stress hormone cortisol, so cycle in carbs smartly to keep things balanced (*Fat Adaptation and Metabolism*).

SECTION TWO: SUPER-CHARGING MITOCHONDRIA

Your mitochondria are like tiny power plants in every cell, making the energy you need to feel great. Let's power them up with some easy tricks.

11.2.1: MITOPHAGY AND BIOGENESIS

Mitophagy is like a spring-clean for your mitochondria—it clears out the old ones so your body can make new, better ones through a process called biogenesis. Activating pathways called AMPK and Sirtuin helps with this cleanup and growth, keeping your energy factories running smoothly. A 36-hour fast can double a marker called LC3, which shows your cells are cleaning house. Urolithin A, found in pomegranates, also kicks off this cleanup—200 mg a day showed promise in early studies. Doing Zone-2 cardio—like a steady jog—while fasted gives you a "double hit," burning fat and boosting this cleanup at the same time (*Mitophagy and Fasting*). A 70-year-old ultrarunner's biopsy showed double the mitochondrial density after following this kind of protocol, proving age is no barrier to powering up. Just wait until new age-reversal technology starts to arrive!

11.2.2: NUTRIENTS AND LIGHT

Certain nutrients and light can give your mitochondria an extra boost, like adding premium fuel to your engine.

- CoQ10 and PQQ work together to make your mitochondria better at churning out energy—think of them as a power-up duo.
- There's a debate about whether Niagen (NR) or NMN is better for boosting NAD+, a key energy molecule, but new research is starting to lean toward NMN for its direct absorption. I take 1,000mg of NMN daily and I've seen outstanding results.
- Using 660 nm red light for 10 minutes a day—like from a small lamp—can raise your ATP energy production by 14%.
- Resveratrol and curcumin act like "stress vitamins," giving your mitochondria a healthy challenge to get stronger.
- Watch out for things like statins or pesticides like glyphosate, which can harm your mitochondria—counter them with a supplement called ubiquinol to keep things running smoothly (*Red Light and ATP*).

If you want to research these and more great supplements, check out my book **Vitamins, Supplements, and Herbs for Health and Longevity: Boost Your Immunity, Increase Energy, and Feel Younger in Minutes a Day.**

11.2.3: EXERCISE PRESCRIPTIONS

Exercise is one of the best ways to keep your mitochondria firing on all cylinders, and you don't need to overdo it.

• Build a base with Zone-2 cardio—where your heart rate is about 180 minus your age—for three sessions a week to make your body better at burning fat.
• Add in six 20-second all-out sprints to boost a protein called PGC-1α, which helps grow new mitochondria.
• Your VO_2 max—how well your body uses oxygen—is a key aging marker, so aim for over 40 ml/kg/min (or around 19-20ml/lb/min) if you're middle-aged to stay in the healthy zone.
• If you're stuck at a desk, try "micro-bursts"—like a 1-minute stair climb—to get the same benefits.
• Every few weeks, take a lighter "deload week" to avoid too much stress from reactive oxygen, keeping your mitochondria happy (*Exercise and Mitochondrial Biogenesis*).

11.2.4: TELOMERASE ACTIVATION: PROTECTING YOUR DNA FOR LONGEVITY

Ever wonder how your cells stay young? Meet telomerase, a superhero enzyme that helps protect the ends of your chromosomes—called telomeres—so they don't wear out too fast. Think of telomeres like the plastic tips on shoelaces; when they fray, your cells can't divide properly, and that speeds up aging. Telomerase keeps those tips strong, and research—like studies from the University of California—shows you can boost it with stuff you're already doing, like exercise (those Zone-2 cardio sessions from Section 11.2.3), meditation (Chapter 14), and even eating foods rich in antioxidants, like berries (Chapter 8). By keeping stress low and moving your body, you're helping telomerase slow down the aging clock, giving your cells more time to keep you healthy and vibrant. For a deep dive in to all of this, check out my book **Stay Healthy, Stay Youthful - The Science of Living to 150: Cutting-Edge Longevity Science for Reversing Aging and Living a Healthier, Longer Life.**

This isn't just lab talk—telomerase activation can add years to your healthspan by keeping your DNA in top shape. A 2013 study found that people who meditated for just 30 minutes a day over 8 weeks had higher telomerase activity, which meant their cells aged slower, tying right into how we've been talking about mindfulness and stress reduction throughout the book. So, when you're doing your daily gratitude jot (Chapter 16) or taking a brisk walk, you're not just feeling good—you're literally helping your body fight aging at the cellular level, one telomere at a time.

SECTION THREE: PRACTICAL METABOLIC RESET

Let's put it all together with a simple reset plan to get your metabolism and mitochondria in top shape, plus some tools to keep it that way for the long haul.

11.3.1: 5-DAY FASTING-MIMICKING DIET

A fasting-mimicking diet (FMD) lets you eat a little while still getting the benefits of fasting—it's like a mini-reboot for your body. For five days, stick to about 800 calories a day, with 10% from protein and mostly plant-based fats—like olives and nuts. On day 6, you'll see a stem-cell rebound in your blood work, which helps your body rebuild and repair. You can DIY it with a grocery list of olives, macadamias, zucchini soup, and kale chips for variety. It's not for everyone—skip it if you're pregnant, have type 1 diabetes, or are frail, and always check with your doctor first. By day 3, your ketones should be over 1.0 mmol/L, showing your body's in cleanup mode (*Fasting-Mimicking Diet Benefits*). While I'm plugging my books... if you're considering weight management, I've used the latest technology to research all of the leading diets in the last century and placed all of the common denominators in my book **The Ultimate AI Diet - Consolidating the Best Diets Over the Last 100 Years: Achieve Fast, Lasting Weight Management with a Personalized, Modern Approach Backed by a Century of Proven Success.**

11.3.2: RE-FEED ALGORITHM

After your FMD, you need to ease back into eating the right way to keep the benefits and avoid issues.

- For the first 24 hours, aim for 30% of your calories from protein—like chicken or tofu—to help your body rebuild without overloading it. Add creatine and electrolytes—like a pinch of salt in water—to prevent puffiness from water retention.
- Slowly bring carbs back over a few days, starting with things like sweet potatoes, to avoid a big insulin spike.
- About 36 hours after your fast, do a strength workout to help your muscles soak up nutrients better.

An auto-immune patient noticed her flare-ups dropped after following this re-feed plan, showing how a gentle restart can calm your system (*Post-Fast Refeeding*).

After a particularly strict diet, I remember tasting the tiniest amount of sugar in a piece of bread and it was overwhelming to me. Give your body a chance to ease back into it after dieting or fasting.

11.3.3: LONG-TERM MONITORING

Keeping your metabolism on track means checking in regularly to see how things are going.

• Every three months, do a "CGM sprint"—wear a continuous glucose monitor for a week to spot any sneaky blood sugar patterns.

• Use a home device to measure your resting metabolic rate (RMR) and see if your calorie burn is shifting over time.

• An AI dashboard can pull together your HRV, glucose, and sleep data to suggest diet tweaks, like adding more protein on active days.

• Once a year, check your epigenetic GrimAge—a test that shows how fast your body's aging—to see how your metabolism is holding up.

Before I continue, I have a personal request for you, my reader…

PLEASE MAKE A DIFFERENCE WITH YOUR BOOK REVIEW

Unlock the Power of Kindness

"Sharing our gifts helps others shine bright."

My father's example as an exemplary healer guided my perception of medicine and health. I was lucky to have great teachers and mentors who helped me grow and pushed me to learn more about enhancing the human condition. Now, I want to help others find their own way towards health and longevity, too.

Would you help someone just like you—excited about getting on the right track but not sure where to start learning how to master health and wellness?

My mission is to educate as many people as possible about new and upcoming age reversal discoveries leading us all towards longer, healthier lives.

But to reach more people, I need your help.

Most people choose books based on reviews. So, I'm asking you to help another by leaving a review.

It doesn't cost anything and takes less than a minute, but it could change someone's journey. Your review could help...

- ...one more person find their way to healthy habits.
- ...one more child know there's a possibility of loving grandparents longer.
- ...one more person gain confidence to adopt a new lifestyle.
- ...one more dream of a better life come true.

If you purchased my book on Amazon, please go into your "Orders" section of your Amazon account and leave your review. It will mean the world to me!

If you love helping others, you're my kind of person. Thank you from the bottom of my heart! **Tad Sisler**

Your engines run clean—now we'll forge resilience with controlled discomfort. That doesn't sound great, but it can do wonders for you! Let's see…

CHAPTER TWELVE
HORMESIS AND RESILIENCE TRAINING

Who wants to be uncomfortable? Maybe you do. Studies show that many "shocks" to our systems can be actually good for us. For instance, regular sauna use cuts heart-disease deaths by nearly 40%—the same heat that forges steel can forge a stronger you! That's the magic of hormesis: small doses of stress can make your body and mind tougher, ready to handle whatever life throws at you.

In this chapter, we'll use heat, cold, and breath to build your physical strength, add some food and fasting tricks once again to keep you resilient, and train your brain to stay cool under pressure—all with a little help from your friends.

SECTION ONE: THERMAL AND HYPOXIC STRESS

Your body loves a challenge, and things like heat, cold, and even holding your breath can make you stronger if you do them the right way. Let's jump into some simple ways to get tougher.

12.1.1: SAUNA SCIENCE AND HEAT-SHOCK PROTEINS

Saunas aren't just for relaxing—they can seriously boost your health by turning on repair proteins called HSP70, which fix up your cells. I've mentioned this already in other chapters but this is relevant to this chapter as well: Sitting in a sauna at 174°F (79°C) for 20 minutes, three times a week, gets those proteins going strong.

That same study I referenced earlier of Finnish men who used saunas four times a week showed a 40% drop in deadly heart events—pretty amazing for just sweating it out! Your blood pressure might dip after, so sip water with a pinch of salt to avoid feeling woozy. If you're older, infrared saunas work great at lower temps, making it easier to handle. After my friend **Larry** had a heart attack, he got his stamina back with a doctor-guided sauna plan, showing how heat can help you bounce back (*Sauna and Heart Health*).

12.1.2: COLD EXPOSURE AND BROWN-FAT ACTIVATION

Cold can wake up your body in a big way—just 11 minutes a week in water or air below 55°F (12°C) can spike your norepinephrine—a stress hormone that gets you alert—by 200%. That chilly dip also fires up brown fat around your neck and chest, which burns sugar to warm you up, giving your metabolism a boost.

Try contrast therapy—sauna then ice—to double that energy kick with a bigger hormone surge. Start with 30-second cold showers and add 15 seconds each week to build up without shivering too much. If you've got heart issues, check with your doctor first to avoid risks like frostbite or heart rhythm problems (*Cold Exposure and Norepinephrine*).

RYAN REYNOLDS

Actor Ryan Reynolds keeps his Deadpool-ready body with trainer **Don Saladino's** functional program—mobility warm-ups, compound lifts spread over five days, and short bursts of cardio—so he stays strong without punishing single-muscle workouts. Recovery is built in: Saladino alternates sauna heat with 2- to 3-minute cold plunges, a ritual he says lets **Reynolds** "rinse and repeat" training sessions while damping soreness.

Reynolds has leaned into the chill himself, telling *GQ* he's "done some of that cold-plunge stuff" to soothe the wear-and-tear of stunt work and even posting a video of a midnight dive into the frigid Adriatic Sea while filming. The science matches the habit: brief cold-water immersion can ease muscle pain, cut inflammation, cool the core, and give a sharp mental jolt—benefits highlighted by sports-medicine experts at Cleveland Clinic.

By pairing approachable icy dips with balanced, evidence-based training, **Reynolds** shows that peak superhero fitness can feel relatable to anyone willing to brave a few shivers. He also uses his cold plunges as a metaphor for life's challenges in general.

"Any kind of crisis can be good. It wakes you up." - Ryan Reynolds

Ryan Reynolds

Credit – Flickr / creativecommons.org

12.1.3: INTERMITTENT HYPOXIA AND BREATHWORK

Breathwork can train your body to handle low oxygen—like being up in the mountains—and make you feel more energized.

• Box breathing—inhale for 4 seconds, hold for 4, exhale for 4, hold for 4—helps you get used to more CO_2, making breathing easier over time.

• Using an altitude mask, which mimics being 3,000 meters up, can boost your red blood cell hormone EPO in just 14 days, helping you carry more oxygen.
• Long breath-holds make your spleen release extra red blood cells, giving you a little energy burst.
• The *Wim Hof* method, with its deep breathing cycles, lifts your mood by raising norepinephrine and dopamine. Always do these seated to avoid passing out—never in water where you could risk a blackout (*Wim Hof and Breathwork*).

SECTION TWO: NUTRITIONAL AND FASTING STRESSORS

Food and fasting can also give your body a healthy challenge, making it stronger from the inside out. Let's look at some easy ways to use them to build resilience.

12.2.1: INTERMITTENT-FASTING VARIANTS

We've described Intermittent fasting (IF) in previous chapters. This is like a mini-bootcamp for your metabolism, and there are a few ways to do it depending on what works for you. Options we've already mentioned like 16:8 (eat in an 8-hour window), 18:6, or 5:2 (two 500-calorie days a week) can fit your schedule—pick one that matches your natural rhythm, like if you're a morning person or a night owl. After about three weeks, your hunger hormone ghrelin settles down, so you'll feel hungry at the same time each day, like clockwork. Sipping black coffee or tea during your fast boosts a cleanup pathway (AMPK), but sweeteners can mess that up, so keep it simple. If you're a woman, a 14:10 window might be gentler to keep your hormones happy. A busy attorney kept his muscle mass using 16:8 fasting and resistance bands, showing you can stay strong even with a packed schedule (*Intermittent Fasting and Ghrelin*).

12.2.2: PHYTOCHEMICAL HORMETINS

Some foods have natural compounds that give your body a tiny nudge to get stronger—kind of like a gentle workout for your cells.
• Sulforaphane in broccoli sprouts kicks on your body's detox system through a pathway called NRF2, helping you clear out toxins.
• Mix curcumin from turmeric with black pepper to make it 20 times easier for your body to use, giving your cells a boost.
• Quercetin, found in apples, paired with zinc, acts like a helper to get zinc into your cells, powering up your immune system.

You want just the right amount of these compounds—too much can be harmful—so stick to normal food portions. Steaming veggies instead of boiling them keeps more of these helpful compounds in your meal (*Sulforaphane and NRF2*).

12.2.3: PROTEIN CYCLING AND mTOR MODULATION

Protein is great, but too much all the time can speed up aging by over-activating the mTOR pathway. Eating leucine-rich foods—like 2.7 grams in a chicken breast—helps build muscle, but you don't want that signal on all the time. As I mentioned previously, try low-protein "Wednesdays," where you cut back to drop your IGF-1 hormone by 15%, slowing down aging without losing strength. On fasting days, have collagen for breakfast—about 10 grams—to support your joints while keeping mTOR low, so your body stays in repair mode. There's ongoing research into micro-doses of rapamycin, a drug that might help with this balance, but it's not ready for everyone yet. An athlete kept setting personal records during a weight cut by cycling protein this way, showing how it can help you stay strong (*mTOR and Protein Cycling*).

SECTION THREE: MENTAL TOUGHNESS AND STRESS INOCULATION

Life can throw curveballs, but you can train your mind to handle them like a champ. Nobody understands mental toughness more than my friend **Mark Redman**, one of the great *Major League Baseball* pitchers of our time. **Mark** achieved his dream of pitching in the World Series. He learned how to cope with stress in big moments like these. **Mark** said:

> ***"It came to the point where you try to do too much out there. You give up a couple of runs and in the back of your head, because of the way things have gone all year, I just tried to battle. It can be wearing on a person, but I've been there before."***

Mark Redman and Tad Sisler

Source – Sisler Private Collection

Let's build some mental toughness with simple habits you can start today.

12.3.1: STOIC PRACTICES FOR MODERN LIFE

The ancient Stoics had some smart tricks to stay calm, and they still work today. Start your morning with "premeditatio malorum"—fancy words for imagining what could go wrong, like a flat tire, so you're ready instead of panicked.

• Choose voluntary discomfort—like a cold shower, speaking in public, or a day with just the basics—to get used to tough situations.
• Use a "dichotomy-of-control" worksheet to focus on what you can change, like your attitude, and let go of what you can't, like traffic jams, to cut down on worry.
• Writing down something you're grateful for each day fights off negative thoughts, keeping your mind balanced.

When I tried to expand my business of 20 years too quickly and I had to shut it down and let go of it, I used Stoic ideas to reframe this scary time in my life, finding calm and clarity to move forward (*Stoic Practices and Resilience*). **Plato** knew the importance of mental toughness when he said:

"We can easily forgive a child who is afraid of the dark; the real tragedy of life is when men are afraid of the light."

Plato
Credit – Chat GPT image generated

12.3.2: RESILIENCE METRICS AND TRACKING

Keeping tabs on your resilience is like checking the oil in your car—it helps you know when to push or pull back.
• Track your morning heart rate variability (HRV), cortisol levels, and mood with an AI diary app to see how you're handling stress.
• If your HRV drops more than 10% below your usual, take a "parasympathetic day" with calming stuff like deep breathing or a walk.
• Each week, fill out a "stress-load" scorecard—rate your sleep, work, and social life on a scale of 1 to 10 to spot what's wearing you out.
• On recovery days, try yoga nidra—a deep relaxation practice—or a more familiar way to completely let go, plus a magnesium supplement and a sauna session to recharge. Over months, you'll see your trend lines show you're getting tougher and better at bouncing back (*HRV and Stress*).

12.3.3: COMMUNITY-BASED CHALLENGES

In the next chapter, I'll delve into the power of community. For one thing, doing tough things with others makes them way more fun—and you're more likely to stick with it.

- Join a cold-dip club or sauna circle to hold each other accountable, cheering each other on as you brave the chill or heat.
- Try a group 5-day fasting-mimicking diet reset, sharing recipes like zucchini soup and checking in to keep everyone motivated—it's easier when you're in it together.
- Gamified leaderboards, where you earn points for sticking to the plan, can boost participation by 40%, making it feel like a friendly competition.
- For safety, do a group medical screening first and use a buddy system, especially for cold dips. Your body and mind are now anti-fragile! (*Group Challenges and Adherence*).

Next I want to emphasize the life-saving power of family, friends, and faith to ground you, keep you stable, and multiply your gains

CHAPTER THIRTEEN
THE ESSENTIAL POWER OF FAMILY AND COMMUNITY

Three months after his second stroke, my 62-year-old father-in-law **Russ** moved into my home with my wife (his daughter) and my four children—within a year, his speech had returned and he was able to reduce his blood thinners, but mostly, his cardiologist cut his blood-pressure meds in half and his fitness level, measured by VO_2 max, rose 22%. His secret weapon wasn't another pill—it was nightly laughter at the dinner table with his kids and grandkids. Family and community aren't just feel-good extras; they're a powerful way to boost your health and live longer. In this chapter, we'll explore how to strengthen your family ties, find purpose through faith and values, build a community that lifts you up, and even tap into the group magic that makes us all stronger.

I heard a radio interview with the Prime Minister of Saudi Arabia. He talked about how Qatar was forced to build a series of nursing homes to qualify for global development standards and goals. The nursing homes are empty because Saudi Arabia's built-in culture, **deeply rooted in religious and tribal traditions, places a strong emphasis on family responsibility for elder care**. It is seen as a moral and religious duty to care for aging parents at home and placing them in institutional care is often considered dishonorable. If families everywhere revered their aging populations, the world would be a better place.

SECTION ONE: FOUNDATIONS OF FAMILY STRENGTH

When my wife **Stephanie** died unexpectedly at the age of 37, my four children and I clung together in grief. Our bond carried us through the next few years and they all became amazing, functional, kind, and compassionate adults. We promised that we would never abandon each other, and that deep connection carried us through some very dark times. Family is like a built-in support squad, and leaning into those bonds can do wonders for your health. Let's look at some easy ways to make your family connections even stronger.

13.1.1: ATTACHMENT CHEMISTRY

Hugs and shared moments with family can work like magic on your body, thanks to the hormone oxytocin that helps you feel connected and calm.

• Just a 12-second hug can bump up your oxytocin levels by about 2 IU/mL, lowering stress hormones like cortisol and inflammation markers like CRP.

• Eating meals together gets your vagus nerve going—the nerve that helps you relax—which can improve your heart rate variability (HRV) overnight, making you feel more rested.

• If your family's far away, try "Family FaceTime Fridays" to video chat and still get that oxytocin boost.

My post-stroke father-in-law saw his blood pressure drop and his speech patterns improve after starting nightly story-time cuddles with his grandkids, showing how simple family moments can heal (*Oxytocin and Family Bonding*). My children would sit with him for hours as he regained his speech and movement, and he was able to impart some wisdom into them and bond so they would never forget their grandfather.

13.1.2: INTER-GENERATIONAL WISDOM LOOPS

When different generations hang out, everyone wins—kids learn from elders, and elders stay active and sharp.

• Grandparents teaching kids music or dance is like a sneaky workout—think of it as high-intensity fun that also keeps their brains young.

• Storytelling, like sharing family traditions around fasting or feasting, can actually pass down healthy habits that protect your genes. A Harvard study found that teens mentored by their grandparents had 18% lower obesity rates, thanks to the active time spent together.
• Gardening with grandkids can add 2,000 extra steps a day for both of you, plus it's a great bonding activity.
• Hosting oral-history nights—where everyone shares old family stories—helps keep your sense of purpose strong and your memory sharp by supporting your brain's hippocampus (*Intergenerational Benefits*).

13.1.3: CONFLICT ALCHEMY

Family fights happen, but you can turn them into opportunities to grow closer. Research by **Dr. John Gottman** shows that families with a 5:1 ratio of positive to negative interactions—like five compliments for every argument—keep their telomeres, the protective caps on your DNA, longer, slowing down aging.
• When tempers flare, try breath-syncing: inhale for 4 seconds, hold for 4, exhale for 6, to calm everyone down in under a minute.
• Set up family "board meetings" to talk about schedules or chores—it can cut household health costs by 22% by reducing stress.
• Write a simple "House Constitution" with your family's values—like kindness or honesty—and rituals, like a weekly game night, to stay on the same page.

My two uncles resolved a long feud by starting a weekly board-game night, turning competition into connection (*Gottman's 5:1 Ratio*).

SECTION TWO:
FAITH, PURPOSE AND MORAL FRAMEWORKS

Faith and a sense of purpose can give you a health boost, whether you're religious or not. Let's see how these ideas can help you live better and longer.

13.2.1: SPIRITUAL NEURO-BIOLOGY

Your brain loves a little spirituality—it's like a calming reset button. Brain scans show that prayer lights up a part called the anterior cingulate, which helps you control impulses and stay calm. A big study in *JAMA* found that going to weekly worship services—like church or a meditation group—can add 4 to 14 years to your life. Singing together in a group, like a gospel choir, releases nitric oxide, a molecule that lowers your blood pressure and helps you breathe easier (Remember, you can get my book VOCAL MASTERY on Amazon to improve your singing skills). Even if you're not religious, secular meditation can boost telomerase—an enzyme that protects your DNA—by 17%, slowing aging. My friend **Gayle** started singing in a gospel choir and reversed her pre-diabetes by singing weekly, finding joy and health in every note (*Spirituality and Longevity*).

13.2.2: MODERN TEN COMMANDMENTS FOR HEALTH

The Ten Commandments aren't just old rules—they can actually guide you to better health in modern life. For instance:

- "Honor your parents" means taking care of elders, which cuts loneliness for everyone—think of it as built-in family support.
- "Remember the Sabbath" reminds you to take a rest day each week, helping your body clock and clearing out sleep debt so you feel refreshed.
- "Do not covet" stops you from chasing endless dopamine hits—like buying stuff you don't need—that can stress you out.

Every three months, check your habits against these commandments to see where you can improve—like cutting back on mindless scrolling.
You'll find a printable worksheet at the end of this chapter to help you track it (*Sabbath and Circadian Health*).

13.2.3: AUTHENTIC ACTION VS. VIRTUE SIGNALING

Living your values—not just talking about them—makes a real difference in your health. When your actions don't match your values, it can raise your stress hormone cortisol by 27%, leaving you on edge. Getting likes on social media for a cause doesn't boost your immune system, but real deeds—like helping a neighbor—do, by increasing your lymphocyte activity.

- Try a 3-step civic micro-action: donate to a cause, volunteer your time, then educate others about it to spread the impact.
- Community gardens can double how much produce people eat, even lowering blood sugar markers like HbA1c in the neighborhood.

A startup pledged 1% of its profits to local shelters, and their team's engagement soared as they felt the impact of their work (*Values-Action Gap and Stress*).

13.2.4: HALF-ANIMAL, HALF-GOD RECONCILING OUR DUALITY

We humans are a fascinating mix—half-animal, half-god—constantly wrestling with our wild instincts and our higher potential.

On one side, our animal selves drive us with raw, primal urges: hunger pushes us to eat, fear makes us run, and stress hormones like cortisol flood our bodies to survive, as we saw with fight-or-flight responses in Chapter 3. But those same instincts can lead us astray—think late-night cravings or snapping at a loved one when stress takes over.

On the other side, our god-self lives in our brain, giving us the power to dream, create, and connect, like when we tap into flow states or feel a deep sense of purpose through family traditions, as we are discussing in this chapter.

This divine spark lets us imagine a better future, solve problems with tools like mindfulness and even sense things beyond the ordinary, like the psychic intuitions we will explore in Chapter 19.

To become the best we can be, we need to reconcile these two halves—honoring our animal instincts while letting our god-self lead the way. That means listening to your body's signals, like using HRV to gauge stress (coming up in Chapter 16), but not letting them run the show—instead, use your brain's higher powers to reframe challenges with optimism (see Chapter 14) or find meaning through community.

Practices like meditation or prayer can quiet your animal instincts, while acts of kindness—like volunteering in a community garden—tap into your god-like ability to connect and create. By balancing these two sides, you'll live a life that's both grounded and inspired, fully embracing your human potential to thrive for years to come.

SECTION THREE: COMMUNITY ARCHITECTURE

Building a strong community around you is like having a health safety net—it keeps you connected and thriving. Let's set up some easy ways to make that happen.

13.3.1: BELONGING HACKS

Feeling like you belong can give your health a big lift, and it starts with small actions.

- Try "3-minute favors"—like grabbing a coffee for a coworker—to spark a web of connection and oxytocin, the bonding hormone.
- Host "Circle-Up dinners" with eight friends and one deep question—like "What's your happiest memory?"—instead of staring at screens, which studies show cuts down on real connection time.
- Start a step challenge on Slack or WhatsApp with friends—it can boost your sticking-with-it rate by 42%, making exercise more fun.
- A buddy system, like pairing up with a friend for workouts, cuts your dropout rate in half.

A corporate team beat metabolic syndrome by taking lunchtime walks together, proving how simple group habits can change your health (*Social Connection and Health*).

"Empathy is the starting point for creating a community and taking action. It's the impetus for creating change." – Max Carver

13.3.2: ENVIRONMENTAL SUPPORT LOOPS

Your neighborhood can help you stay healthy without much effort if it's set up right. Living in a "walk-score" area of 70 or higher—where you can easily walk to shops or parks—adds 1,200 steps a day without even trying. Pocket parks in your area can lower neighborhood blood pressure just by giving people a green space to relax. Benches every 100 meters encourage quick chats with neighbors, boosting your social life on the go. Areas zoned for farmers' markets see a 10% jump in fruit and veggie purchases, making healthy eating easier. There's a community-lobby checklist at the end of this chapter to help you push for these changes in your town (*Walkability and Health*).

13.3.3: MEASURING SOCIAL VITAL SIGNS

You can track how connected you feel to see how it's helping your health—it's like taking your social temperature.

• Wearables can measure the tone of your voice to spot stress, showing which days you're feeling more connected.

• Keep a weekly "Connection Score" journal—rate how close you felt to others from 0 to 10, maybe after a coffee date or family call.

• Check your HRV after meeting friends at a café; a higher number means those hangouts are recharging you.

• Loneliness raises a marker called IL-6, but volunteering can bring it down—the more hours you give, the better you'll feel. There's a mini-exercise worksheet at the end to map your five closest contacts, rate how often they send you "positive-health signals"—like encouraging a walk—then design a weekly group ritual, like a Sunday hike, to keep those vibes strong (*Loneliness and IL-6*).

My mother equated being alone with loneliness. I felt so bad for her, yet I resolved to equate being alone with solitude and reflection, allowing myself to be my own best friend. Eventually, my mother went out and volunteered and her loneliness ebbed. You can always reframe your situation in a positive way if you try.

SECTION FOUR: SWARM-HEALTH AND BEHAVIORS

Just like animals work together to survive, humans can boost their health by syncing up with their community. Let's see how group habits can make you stronger.

13.4.1: FRIEND GROUPS

A 12-year Harvard study in Framingham found that habits like obesity, smoking, and even happiness spread through friend groups in patterns that match how ants make decisions—little signals add up to big changes when enough people join in. If your friends start eating healthier, you're more likely to do the same, kind of like a health domino effect (*Framingham Social Networks*). My four adult children frequently create their own exercise or weight management challenges to keep each other in check.

13.4.2: STADIUM SYNC

I've talked about this but it's worthy of another mention in a new context. Ever been to a huge concert or game and felt totally in sync with the crowd? Research from Ulm University shows that when 50,000 fans breathe and chant together—like at a **Taylor Swift** concert—their heart rhythms and brain waves start to match, boosting everyone's vagal tone, which helps you relax and feel good (*Stadium Sync Study*).

13.4.3: GROUP-FLOW BOOSTS

When you're in a group that's totally in the zone—like a basketball team on a hot streak or an improv comedy troupe—your body gets a health boost too. A 2021 study from Oxford's Social Neuroscience Lab found that group flow, where everyone's vibing together, can actually turn on immune-boosting genes, making you stronger just by being part of the team (*Group Flow and Immunity*). The trick is to keep the group in the zone!

My friend **Tony Kaye** knew a lot about group dynamics. He was the legendary **David Bowie** and **Moody Blues** keyboardist... and founding member of the band **"Yes".** We talked about different groups we've been in as musicians. Some were amazing, with talented people working together with 'one mind' and great attitudes. Other groups were not so easy to deal with. One band member might be unhappy or have too big of an ego and it ruined the experience. No matter how talented you are, if you can't work with others, you're useless. But, there's always a lesson to be learned in working within a group, and the benefits usually outweigh the disappointments if you can figure out a way to stay out of the politics of the situation. **Tony** summed it up best when he said:

"Everything that happens is meant to be. It's meant to happen like that. But sometimes you don't know at the time that it's meant to be a disaster."

Tony Kaye and Tad Sisler
Source – Sisler Private Collection

Attitudes and emotions are everything. They will make or break you. So, now that your village powers your physiology, let's talk about these attitudes and emotions… the inner dialogue that programs every one of your cells.

CHAPTER FOURTEEN
ATTITUDES, ACTIONS AND EMOTIONAL MASTERY

A review of 1.7 million people found pessimists die 7.2 years earlier than optimists—even when diet and exercise are identical. That's a big deal—it means your mindset can literally add years to your life! Read this book I wrote to get the full picture: **The Science of Positive Thinking: How Mindset, Daily Habits, and Emotional Well-being Can Add Years to Your Life.** In this chapter, we'll explore how your brain handles emotions, show you how to use feelings to boost your health, and give you a toolkit to master your mind so you can live longer and happier.

SECTION ONE: NEURO-BIOLOGY OF OPTIMISM

Your brain is like a control center for your emotions, and understanding how it works can help you stay positive and strong. Let's break down the science of optimism in a way that's easy to get.

14.1.1: TRIUNE BRAIN 2.0

Your brain has three main parts that handle emotions, kind of like a team working together—or sometimes fighting! This very much ties into the half-animal, half-god concept we were discussing in Chapter 13.

• The reptilian brain, the oldest part, kicks in with a fight-or-flight response, dumping sugar into your blood to get you moving fast when you're stressed.

• The limbic brain can get stuck in "doom loops," where your fear center—the amygdala—keeps you worried, but calming techniques like deep breathing can reset it.
• Your neocortex, the thinking part, can reframe tough situations—using tools like cognitive behavioral therapy (CBT) to shrink those fear circuits over time.

When you reflect on things you're grateful for, brain scans show a boost in calming alpha waves, helping you feel more at peace. A paratrooper used a simple breath-counting trick—inhaling for 4, exhaling for 4—to steady his nerves mid-combat, showing how you can take control even in high-stress moments (*CBT and Amygdala*).

14.1.2: CHOICE-POINT MICRO-SECONDS

Viktor Frankl, a famous thinker, said there's a tiny space between something happening and how you react—that's where your freedom lies. An emotional surge, like getting mad, lasts less than 90 seconds unless you keep thinking about it and fuel the fire.
• Try "name it to tame it"—say out loud, ***"I'm frustrated,"*** to shift control to your thinking brain and calm down faster.
• Keep a small object, like a coin, in your pocket or repeat a calming phrase—like ***"I've got this"***—to snap you out of stress and make better choices.
• HRV rings can even buzz when your stress levels get too high, giving you a real-time nudge to pause and breathe (*Emotional Surge Duration*).

"Between stimulus and response there is a space. In that space is our power to choose our response. In our response lies our growth and our freedom."
- Victor Frankl

Victor Frankl
Credit – Store norske leksikon / creativecommons.org

14.1.3: VALUES-DRIVEN DECISION MATRIX

Living by your values keeps your brain's happy chemicals—like dopamine and serotonin—in balance, making you feel good about your choices.

- Write a personal mission statement, like ***"I want to spread kindness,"*** to guide your decisions.
- Use IF-THEN scripts to make tough choices easier—like, ***"IF I'm tempted to skip my workout, THEN I'll do just 5 minutes."***
- Each week, check your calendar: does how you spend your time match what matters to you, like family or health?
- Turn it into a game—give yourself micro-points for actions that line up with your values, like helping a friend.

A start-up CEO dodged burnout by using a "value triage board" to focus on what really mattered, keeping her energy and focus on track (*Values and Dopamine*).

SECTION TWO: HARNESSING EMOTIONS FOR LONGEVITY

Your emotions aren't just feelings—they can actually help you live longer by boosting your body's defenses. Let's see how to turn your emotions into a health superpower.

14.2.1: POSITIVE AFFECT AND IMMUNITY

Feeling good does wonders for your immune system—like giving it a big hug from the inside. Laughing can boost your natural killer (NK) cells, which fight off sickness, for two hours after a good chuckle. Oxytocin, the bonding hormone you get from hugs, speeds up wound healing in both mice and humans, helping you recover faster.

Writing down three good things each day—like a sunny day or a kind word—can lower your evening stress hormone cortisol by 25%, helping you relax. Listening to music, like a soothing playlist, can balance your immune system's inflammation markers, keeping you healthier. A cancer patient credited weekly comedy club visits with cutting her pain meds, showing how laughter can be a real healer (*Laughter and NK Cells*).

Much of optimism is reframing a situation and finding the positive element in it... and then, move on. **Edgar Cayce** said:

> ***"There is progress whether ye are going forward or backward! The thing is to move!"* ... he also said:**
> ***"It is thought and feeling which guides the universe, not deeds."***

Remember, thoughts are things and your attitudes and emotions can shape your destiny.

Edgar Cayce

Credit – Chat GPT Image Generated

14.2.2: STRESS ALCHEMY—EUSTRESS VS. DISTRESS

Remember, stress isn't all bad—it's like a U-shaped curve where too little means you're not growing, but too much can make you sick, so the sweet spot is key.

- Use 4-7-8 breathing—inhale for 4 seconds, hold for 7, exhale for 8—to drop your blood pressure in just 90 seconds when you're feeling overwhelmed.
- Combining sauna sessions with a cold plunge can slash your dementia risk by 30%, turning stress into a brain-protecting tool, as we saw in Chapters 11 and 12.
- Make a "recovery assets" list—things like good sleep, stretching, or playing with friends—to help you bounce back after tough days. We've talked about hormesis before, and this is another way to use it to make you stronger (*Sauna and Dementia Risk*).

14.2.3: EMOTIONAL TRANSMUTATION

Napoleon Hill, already quoted in this book, talked about turning negative emotions into positive action, and modern brain scans show how this works by linking emotions to new thoughts.

- Turn envy into inspiration with mental contrasting—use the WOOP method (Wish, Outcome, Obstacle, Plan) to imagine your goal, see the challenge, and plan to overcome it.
- EFT tapping, where you tap on specific body points while talking through your feelings, can boost your calming vagal tone and cut anxiety scores by 40%.
- Channel anger into a quick HIIT sprint—use that adrenaline to power a 20-second burst instead of stewing.

A writer turned her grief into a bestselling memoir, and her health markers—like stress and inflammation—improved as she poured her emotions into her work (*EFT and Anxiety*).

My friend, actress and influencer **Khloe Kardashian**, is a master of turning negative emotions into positive ones. She said:

"You can't expect everyone to love you. I'm not someone who just wants to throw out hate, just because. I know my soul is beautiful; I know I'm a good person. And that will never change for me."

Tad Sisler with Khloe Kardashian and Robin Dougan

Source – Sisler Private Collection

14.2.4: SOUND AND VIBRATION: HEALING THROUGH FREQUENCIES

Ever thought sound could heal you? Sound and vibration—like from binaural beats or those cool Tibetan singing bowls—can do wonders for your body and mind, like a mini-vibration massage for your cells. These frequencies can shift your brain waves into a calmer state, like when you're in deep meditation (Chapter 14, Section 14.3.1), lowering stress hormones like cortisol and firing up your parasympathetic nervous system—that's the "rest and digest" mode we talked about in Chapter 12.

A 2022 study found that just 20 minutes of sound therapy with low-frequency vibrations can improve sleep and reduce pain, helping your body repair itself faster, which is key for living longer. You can try this at home—put on some binaural beats (there are free tracks online) or even hum to yourself to feel the vibes in your chest, which can boost your mood and calm your nerves, just like music therapy we mentioned earlier (Section 14.2.1). By adding sound and vibration to your routine, you're giving your body another way to heal naturally, reducing inflammation and stress while keeping your emotions in check, setting you up for a healthier, longer life with every soothing note.

SECTION THREE: PRACTICAL MIND-MASTERY TOOLKIT

Let's wrap up with some easy tools to keep your mind sharp and your emotions in check, so you can handle anything life throws at you.

14.3.1: MINDFULNESS 3.0

Mindfulness helps you stay present, even in our distraction-filled world, and there are new ways to make it work for you.

- Try VR-assisted mindfulness sessions—like a virtual forest walk—to train your brain to focus better despite all the noise.
- Between emails, take a quick 10-breath micro-meditation—inhale deeply, exhale slowly—to reset your mind in seconds.
- Brain scans show focused attention quiets the part of your brain that wanders, keeping you grounded.
- Compassion meditation, where you wish kindness for yourself and others, can boost telomerase—an anti-aging enzyme—by 17%, protecting your cells.

A teen athlete used a VR cave to practice mindfulness, helping him nail his free-throw focus and win games (*Compassion Meditation and Telomerase*).

14.3.2: NEURO-NUTRIENTS AND ADAPTOGENS

Certain nutrients can give your brain a little extra love, keeping you calm and focused.

- Mix L-theanine with caffeine in a 2:1 ratio—like in green tea—to stay alert without the jitters, perfect for a busy day.
- Ashwagandha, at 600 mg a day, can lower your stress hormone cortisol by 23%, helping you chill out naturally.
- A higher omega-3 index—think fish oil or salmon—keeps your brain's memory center, the hippocampus, in good shape.
- DNA-guided nootropic stacks let you personalize nutrients like choline and B-vitamins for your brain's needs, but cycle them monthly to avoid your body getting too used to them.

Celebrities like **Gwyneth Paltrow** use adaptogens to stay sharp during intense filming schedules, showing how these can fit into a busy life (Ashwagandha and Cortisol). Don't forget to check out my book **Vitamins, Supplements, and Herbs for Health and Longevity: Boost Your Immunity, Increase Energy, and Feel Younger in Minutes a Day.**

14.3.3: DIGITAL HYGIENE AND COGNITIVE RESERVE

Your digital habits can either drain or boost your brain, so let's clean them up.

- Try a dopamine-fast evening—no social media after 8 p.m.—to give your brain a break from constant scrolling.

• Use blue-light filters on your screens to protect melatonin, the sleep hormone that helps your brain store memories while you rest (*Blue Light and Sleep*).
• "Inbox-Zero Fridays," where you clear your email by the end of the week, can raise your Monday HRV by 5 milliseconds, starting your week calmer.
• Keep your brain sharp by rotating skills—like learning a new instrument, language, or even coding—think of it as a gym for your mind. Playing piano has been one of the greatest joys and blessings of my life.

Now that we've upgraded your inner software, let's tap into quantum tech and AI to super-charge the hardware.

CHAPTER FIFTEEN
QUANTUM TECH PLUS AI
ROCKET FUEL FOR HUMAN POTENTIAL

Our world is about to be rocked with quantum computing. If today's fastest laptop is a bicycle, a mature quantum computer is a rocket ship—able to solve molecular problems in minutes that would take regular machines 10,000 years! That kind of power is changing how we understand our bodies, from the tiniest cells to our daily health habits. In this chapter, we'll break down how quantum tech and AI can boost your potential, helping you eat better, think sharper, and even live longer—all in ways that feel like having a super-smart buddy by your side.

SECTION ONE: QUANTUM 101: WHY SUB-ATOMIC MATH MATTERS TO YOUR CELLS

Quantum tech might sound like sci-fi, but it's really just a super-fast way to do math that helps us understand your body better. Let's see how it can work for you.

15.1.1: QUBITS AND BIOLOGICAL MODELING

Quantum computers use something called qubits, which are like magic puzzle pieces that can try millions of solutions at once—way faster than regular computers. A company called D-Wave used 2,000 qubits to figure out how proteins fold 200 times quicker than old-school machines, which could lead to better medicines (*D-Wave Protein Folding*).

Scientists think some tiny reactions in your body, like how enzymes work, might use quantum tricks to happen faster—quantum tech can help us copy that to make cleaner hormones for your health.

The Buck Institute, a big research group, used IBM's quantum cloud to test molecules that clear out old cells, speeding up anti-aging research. Good news: you can now use quantum cloud services for less than $10 a minute to run basic health simulations, and that price will go down dramatically soon (*Quantum Computing in Biology*).

15.1.2: PRECISION NUTRITION VIA QUANTUM METABOLOMICS

Quantum tech can help you eat smarter by predicting how your body reacts to food—like a crystal ball for your meals. It can guess your blood sugar after eating with 94% accuracy, way better than older methods at 72%, so you know exactly what keeps you steady. It can even make a "digital twin" of your gut microbes, figuring out how much good stuff—like short-chain fatty acids—they'll make from any recipe you try. In the future, it might simulate how your liver cleans out supplements before you take them, so you know what's safe.

Your data stays private with something called federated learning, which only shares the math, not your personal info. The result? A grocery list tailored to boost your energy molecule NAD+, helping you feel more vibrant (*Quantum Metabolomics*).

15.1.3: ROAD MAP TO THE 2030s

This book was completed in 2025. At this writing, quantum tech is moving fast, and here's what's coming down the road to help your health.

- By 2026, new photonic qubits will work at room temperature, so hospitals can have quantum chips right at your bedside.
- In 2027, quantum tools will make CRISPR gene editing 90% faster by spotting mistakes early.
- By 2028, a public project called Rejuvenome DAO will let people team up online to design quantum-powered drugs.
- In 2029, we might see a single pill that clears out old cells in Phase I trials, thanks to quantum research.

• And by 2030, you could have a tiny quantum device at home that checks your gut microbes in real time, giving you instant health tips (*Quantum Tech Roadmap*). But, since quantum computing is moving so fast, we may have literally thousands of other remedies, treatments, and cures for diseases by then.

SECTION TWO: AI AS YOUR 24/7 HEALTH CO-PILOT

AI is like having a health buddy who's always there, watching your stats and giving you tips to stay on track. Let's see how it can help you every day.

15.2.1: WEARABLES MEET LARGE-LANGUAGE MODELS

Your smartwatch isn't just for counting steps anymore—it can team up with AI to keep you healthy. Advanced AI, like the kind behind chatbots, can now read 100,000 heartbeats a day and spot irregular rhythms before you even feel them, catching things like arrhythmias early. If your continuous glucose monitor (CGM) shows your blood sugar creeping above 110 mg/dL, the AI can suggest a quick meal swap—like swapping chips for nuts—to keep it steady. The AI runs right on your phone, so your data doesn't go to the cloud, keeping it safe from leaks. It can even listen to 15 seconds of your voice and detect if you're feeling down, helping you catch depression early. A retiree avoided an ER visit when his smartwatch AI warned him of a silent heart rhythm issue, getting him to the doctor just in time (*AI and Wearables*).

15.2.2: PREDICT→PREVENT→REVERSE LOOP

AI can look at your daily habits—like sleep, steps, and heart rate—to predict health risks before they get serious, kind of like a weather forecast for your body. It can spot "pre-frailty"—a sign you might struggle later—18 months before doctors would notice, giving you time to act.

It uses fun nudges, like earning points for a 3-day streak of healthy habits, which can boost your stick-to-it rate by 40%.

If life throws a curveball—like a new baby—the AI adjusts your health plan overnight to keep you on track. Every three months, you'll get a PDF health forecast to share with your doctor, making check-ups a breeze (*AI Predictive Health*).

15.2.3: ETHICS, BIAS AND DATA GUARDRAILS

AI can be amazing, but it's got to be fair and safe—here's how to make sure it works for everyone. Some AI models might miss needs for certain groups, like minorities, so always ask for "model cards" that show how the AI was built to be fair.

A trick called differential privacy keeps your unique DNA data safe, so no one can misuse it.

The World Health Organization sets rules for AI, like making sure it explains its advice, gets your okay to use your data, and has an emergency off switch if something goes wrong. You can store your info in a personal "data vault" and let apps use it with temporary tokens that expire, keeping you in control.

A rural clinic used an open-source AI model to cut stroke risk by 23% in their community, showing how fair AI can make a big difference (*AI Ethics in Health*).

SECTION THREE: BRAIN-COMPUTER INTERFACES AND HUMAN ENHANCEMENT

Brain-computer interfaces (BCIs) sound futuristic, but they're already here, helping people think better and move easier. Recently, I watched a video of a paraplegic standing up and walking with a device attached to his torso and legs along with a BCI. Let's explore how they can boost your potential.

15.3.1: FROM PROSTHETICS TO MEMORY BOOST

BCIs, like those from Neuralink and Synchron, are tiny devices that connect your brain to tech—they can now read your thoughts fast enough to type 90 words per minute just by imagining handwriting. Early tests showed a hippocampal chip—a memory booster—helped people recall 37% more words, like remembering a grocery list easier. There are risks, like brain scarring or hacking, so companies are adding safety layers, like RFID firewall gloves to block bad signals.

By 2026, non-invasive ultrasound BCIs will let you use them without surgery, making them safer. The FDA is fast-tracking these devices to get them to people quicker (*BCI and Memory*).

15.3.2: FLOW STATE ON DEMAND

I love being "in the zone," where everything just clicks! BCIs can help you get there faster. Using EEG headsets with tiny vibrations, they guide your brain into a calm, focused state called the alpha-theta bridge, perfect for peak performance. Athletes in a 2024 study boosted their reaction speed by 15% after six sessions, like hitting a tennis ball quicker. By 2027, portable ear-clip devices will cost under $300, making this tech easy to use at home.

Some folks debate if it's a shortcut compared to building skills the old way, but it's like training wheels for your brain. An e-sports veteran extended his career by three years using a flow-BCI module, staying sharp in high-stakes games (*Flow State and BCI*).

"Hopeful thinking can get you out of your fear zone and into your appreciation zone." – Martha Beck

15.3.3: AUGMENTING MORAL AND SOCIAL INTELLIGENCE

BCIs can even make you a better person by helping you connect with others. They can spot when you're not feeling empathetic during a chat—like at a work meeting—and send a little buzz to your wristband to remind you to listen better (my wife would have a heyday with me with this one!). Early trials with a neuro-patch that mimics oxytocin—the trust hormone—boosted trust scores by 18%, helping people get along better. A "digital conscience" overlay can flag harsh words before you hit send online, so you don't regret that snarky comment.

But there's a big question: are we outsourcing our kindness to tech? It's something to think about as these tools grow. There's a study called "From Ant Trails to Quantum Trails," showing how ant teamwork inspires Google's DeepMind AI for protein folding and robot surgeons at Johns Hopkins to plan better paths.

Now that your village powers your physiology, we'll master the inner dialogue that programs every cell.

CHAPTER SIXTEEN

DESIGN YOUR 90-DAY LONGEVITY BLUEPRINT

A wise old man once asked me that if I had just one car to last a lifetime, would I take care of it differently than I do my existing car? Fortunately, we can trade in our automobiles at some point, but we only have this one body, so let's make the best of it. If you treated your body like a start-up, what 90-day sprint would give the highest return on lifespan?

Just like a business needs a focused plan to grow, your health can thrive with a clear, step-by-step strategy to add years to your life. In this chapter, we'll assess where you're at, craft a 90-day plan to boost your health, and set you up to keep it going strong—no matter what life throws at you.

I WANT TO OFFER YOU A FREE GIFT

I hope you're loving this book so far. In my **HEALTH AND LONGEVITY SERIES,** I address slowing aging, vitamins and supplements for longevity, an excellent AI diet and weight management plan, how positive thinking can add years to your life, and the power of your mind and body (this book). Before we get into crafting your own personalized 90-day longevity blueprint, I've also created a **TWELVE-STEP ACTION PLAN FOR LONGEVITY AND HEALTHSPAN**, a roadmap for health and longevity encompassing elements from **all** my books, and I want to share it with you.

If you want a free copy of my plan, email us at...
<< **modernrenaissancepublishing@gmail.com** >>
with the subject line **12-STEP ACTION PLAN FOR LONGEVITY,** and I'll email you back a free copy at no obligation whatsoever to you as a heartfelt thanks for reading this book. Or you can access it through our website at **https://www.modernrenaissancepublishing.com.**

SECTION ONE: ASSESS AND BASELINE

Before you start any plan, you need to know where you're standing—think of it like taking a quick health selfie to see what's working and what needs a boost. Since there is no "one size fits all" plan, let's get a clear picture of your starting point.

16.1.1: CORE METRICS SNAPSHOT

First, let's grab some key numbers to see how your body's doing—it's like checking the dashboard of your car before a road trip.

- Get a lab panel with tests like a complete blood count (CBC), metabolic panel (CMP), ApoB for heart health, fasting insulin, hs-CRP for inflammation, vitamin D, and a DNAmAge test to check your biological age.
- Measure your vitals: your waist-to-height ratio (waist should be less than half your height), grip strength with a handheld device, and your heart rate variability (HRV) average over seven days to see how well you handle stress.
- Track your sleep with a wearable or journal—how many hours, and how rested do you feel? Rate your mood and energy on a 1-to-10 scale each day.

My friend, a nurse named **Maria** found a hidden insulin spike through these tests, even though her BMI was normal, helping her catch a problem early (*Lab Metrics for Longevity*).

"Inaction breeds doubt and fear. Action breeds confidence and courage. If you want to conquer fear, do not sit home and think about it. Go out and get busy." – Dale Carnegie

16.1.2: LIFESTYLE INVENTORY

Now, let's look at your daily habits—like taking a quick peek into your life's toolbox.

- Do a 24-hour recall: write down everything you ate, how much you moved, your screen time, and who you hung out with.
- Map out your stress—what's weighing on you from work, family, finances, or your sense of purpose?
- Check your tech setup: what wearables or apps are you using, like a smartwatch or food tracker, and where are the gaps, like missing sleep data?
- Rate your habits as red (needs work), yellow (so-so), or green (rocking it) for a quick visual—like marking your diet red if you're eating too many snacks.
- Pick a "keystone habit"—the one that pulls other areas up or down, like late-night scrolling messing with your sleep (*Lifestyle Inventory and Health*).

16.1.3: PURPOSE AND MOTIVATION SCAN

Your "why" is the fuel for your plan, so let's find what drives you.

- Write a one-sentence North-Star Goal—like *"I want to dance at my granddaughter's wedding"*—to keep your eyes on the prize.
- Use an Ikigai diagram to list what you love, what you're good at, what your community needs, and what you can get paid for, helping you find your sweet spot. An **Ikigai diagram** is a four-circle Venn diagram popularized in the West to help people visualise where their deepest motivations and practical realities overlap. While "ikigai" (生き甲斐) in Japanese simply means "that which makes life worth living," the diagram breaks this broad idea into four intersecting questions: What do you love (what activities light you up)? What are you good at (where do your skills shine)? What does the world need (how can I serve or contribute)? What can I be paid for (what value will people reward)?

• Try a values card sort—a free online tool—to figure out what matters most, like family or growth.
• Rate your readiness to start this plan from 1 to 10; if it's below 7, find an accountability partner, like a friend to check in with.
• Do a quick 5-minute exercise: record a visualization of your "future self"—imagine how you'll feel in 90 days, full of energy and joy (*Ikigai and Purpose*).

SECTION TWO: CRAFTING THE 3-RING ACTION PLAN

Now let's build your 90-day plan with three "rings" of action—daily habits, weekly boosts, and quarterly check-ins—to keep you moving forward. It's like setting up a simple routine to grow your health step by step. The trick is not to make it look like a 3-ring circus!

16.2.1: DAILY CORE (15 MINUTES EACH RING)

Your daily core is like the heartbeat of your plan—small, quick actions that add up big.
• For your mind, spend 60 seconds on box breathing—inhale for 4, hold for 4, exhale for 4, hold for 4—then jot down something you're grateful for, like a sunny morning.
• For your body, do a movement snack: try door-frame push-ups or sun salutes for a quick stretch and strength boost.
• For recovery, take a 2-minute cold rinse in the shower and start a digital sunset—turn off screens an hour before bed to help you sleep better.

Stack these with habits you already have, like doing push-ups while your coffee brews or breathing during lunch. Track your streaks in a habit app, aiming for 80% consistency to make it stick (*Box Breathing and Stress*).

16.2.2: WEEKLY UP-SHIFT

Each week, add some bigger moves to level up your health—like giving your plan a little turbo boost.

• Do two Zone-2 cardio sessions—like a brisk walk where you can still talk—for heart health, two strength workouts, and one sauna or cold-plunge session to build resilience, as we covered in Chapter 12.
• Spend one 120-minute block in green space—like a hike or gardening—to soak in nature's benefits.
• On Sundays, meal-prep a Mediterranean-plus menu—think lots of veggies, olive oil, and fish—to set yourself up for the week.
• Plan a family or friend dinner to boost oxytocin, the bonding hormone, and keep you accountable.

My free-wheeling friend **Paula** dropped her blood pressure after six straight "nature Sundays," showing how weekly outdoor time can work wonders (*Nature and Blood Pressure*).

16.2.3: QUARTERLY RE-CODE

Every three months, let's tweak your plan to keep it fresh and effective—like updating your phone to the latest version.

• Test your microbiome with a kit to see how your gut's doing, then adjust your fiber and fermented foods—like adding more kimchi—based on the results.
• Retest your epigenetic clock to see how your biological age is changing, and tweak your fasting or sauna frequency if needed.
• Do a 48-hour fasting-mimicking diet or protein-cycling reboot, like we talked about in Chapter 11, to reset your system.
• Spend a month swapping skills—learn a new language or instrument—to keep your brain sharp and growing.
• Review your North-Star Goal and adjust your roadmap to stay on track for that big dream (*Microbiome Testing and Health*).

SECTION THREE: SUSTAIN, ITERATE, OVERCOME OBSTACLES

You've got your plan—now let's make sure it sticks, tweak it as you go, and handle any bumps in the road. It's like keeping your start-up running smoothly for the long haul.

16.3.1: METRICS AND DASHBOARDS

Checking your progress keeps you motivated—like seeing your business grow with each report.

- Spend 5 minutes each week reviewing your weight trend, HRV, and mood graph to spot what's working.
- An AI health-bot can turn your raw data into simple wins, like "Great job—your sleep score's up!" Use color-coded goals: green means you're on target, yellow means watch out, and red means it's time to adjust.
- Celebrate micro-wins—like hitting your step goal—to get a dopamine boost that makes habits stick.
- Share a screenshot of your dashboard with your accountability buddy to keep you both cheering each other on (*Tracking Metrics for Health*).

16.3.2: SOCIAL AND ENVIRONMENTAL REINFORCEMENTS

Your friends and surroundings can help you stay on track—like having a great team at your start-up.

- Create a "Longevity channel" in your family or group chat to share tips, like a new recipe or a walking route.
- Host a monthly potluck with your blueprint recipes—think Mediterranean salads—to make healthy eating a fun group thing.
- Swap your desk chair for a stability ball to sneak in some core work while you sit.
- At work, push for walk-meetings, add plants for a nature vibe, and use circadian lighting to feel better all day.

A team of nurses started a cold-plunge club, using an HRV leaderboard to spark friendly rivalry, making resilience training a blast (*Social Support and Health*).

16.3.3: PLAN B FOR SETBACKS

My friend, legendary actress **Dyan Cannon** had amazing success, but she also dreaded the inevitable setback. **Dyan** said:

"Have you noticed when you start getting happy, you say, uh-uh, I'd better watch out. I feel too good. Something's going to happen."

Dyan Cannon and Tad Sisler
Source: Tad Sisler's Personal Collection

Life can throw curveballs, so you need a backup plan to keep going—like a smart start-up pivoting when things get tough.

- If you're sick for a week, switch to gentle stuff: do breathwork, visualize your goals, and sip protein soup to stay nourished.
- For travel, pack a hack kit: resistance bands for quick workouts, fasting to simplify meals, a noise-masker for sleep, and probiotic sachets to keep your gut happy.
- If your motivation dips, reread your "why"—that North-Star Goal—and a letter to your future self to get fired up again.
- If your epigenetic clock doesn't budge after two cycles, re-baseline with new tests to adjust your plan (*Resilience and Setbacks*).

Blueprint loaded—now, let's stress-test it against real-world hurdles and future tech in Chapters 17 & 18 before our final launch sequence.

CHAPTER SEVENTEEN
STRESS-TESTING YOUR BLUEPRINT
PLATEAUS, PIVOTS, AND REAL-WORLD CHAOS

Your plan is only as good as its first collision with reality—yet every collision is a data-point, not a dead-end! Life doesn't care about your perfect 90-day blueprint; it'll throw curveballs like travel, sickness, or stress to test you.

In this chapter, we'll tackle those external disruptions, push through internal roadblocks, and learn how to adjust your plan so you keep moving forward—no matter what.

SECTION ONE: EXTERNAL DISRUPTORS

Life can mess with your plan in big ways, like travel or getting sick, but you've got tools to stay on track. Let's handle these outside challenges like a pro.

17.1.1: JET LAG, SHIFT-WORK AND CLOCK CHAOS

Traveling across time zones or working odd hours can throw your body clock out of whack, but you can bounce back fast.

• If you're flying west to east, you'll lose some deep REM sleep, so try to get an extra 90 minutes of shut-eye before your flight to bank some rest.

• Once you land, eat meals on the new time zone's schedule—it can reset your gut clock in just 48 hours.

• Use a 10,000-lux light box, about as bright as a sunny day, at your "new" morning time to shift your body clock 30 minutes earlier each day.

• A tiny 0.5 mg dose of melatonin works as well as a bigger one to adjust your sleep cycle, but leaves you less groggy.

• An ICU nurse kept her HRV steady despite flipping her schedule weekly, using these tricks to stay balanced (*Jet Lag and Circadian Reset*).

You can always adopt the philosophy of my old friend, legendary multi-platinum recording artist **Glen Campbell. Glen** said:

"I like to start the day early, it keeps me out of trouble."

Glen Campbell Performing with Tad Sisler

Source – Sisler Private Collection

17.1.2: ACUTE ILLNESS OR INJURY

Getting sick or hurt can pause your plan, but you can use it as a chance to heal smarter. A fever actually ramps up repair proteins in your body—help them out by drinking extra water and taking glycine, a supplement that supports recovery.

After surgery, a fasting-mimicking diet can kickstart autophagy—your body's cleanup mode—to speed up tissue repair, as we saw in Chapter 11. If you're injured, blood-flow-restriction (BFR) bands let you keep muscle strength with just 30% of your usual workout weight, so you don't lose progress.

While you're resting, visualize your movements—like lifting weights—to keep your brain's motor skills sharp. For the first 72 hours, take an anti-inflammatory mix of omega-3, curcumin, and vitamin C to help your body heal (*Fasting and Autophagy*).

17.1.3: LIFE TRANSITIONS (BABY, JOB LOSS, CAREGIVING)

Big life changes—like having a baby, losing a job, or caregiving—can turn your routine upside down, but you can adapt.

• If your sleep's all over the place, aim for two 3-hour blocks of rest plus 20-minute naps to get through the day.
• Make a "stress budget" sheet: list your top five worries—like finances or lack of sleep—and pick one small action for each, like setting a budget or asking for help.
• Sneak in 5-minute micro-workouts—like a quick jog—to keep your fitness level up when you can't hit the gym.
• Write a script to ask your community for help, turning your to-do list into a shared one—like getting a friend to cook a meal.

A new father kept his blood sugar steady by taking stroller walks and sipping protein shakes, showing how to fit health into a busy life (*Stress and Life Transitions*).

SECTION TWO: INTERNAL RESISTANCE

Sometimes the biggest hurdles come from inside—like feeling bored or stressed—but you can push through with some simple tricks. Let's tackle those inner struggles.

17.2.1: HABIT FATIGUE AND BOREDOM

My friend, legendary multi-platinum recording artist **Kenny Rogers** said:

"There is a trade off - as you grow older you gain wisdom but you lose spontaneity."

Tad Sisler and Family with Kenny Rogers

Source – Sisler Private Collection

But it doesn't have to be that way! Doing the same thing over and over can get old, but a little variety can keep you excited.

- Switch up your exercise every four weeks—like trading running for swimming—to give your brain a dopamine boost and make it fun again.
- Add tiny new things to your routine, like a new spice in your meals, walking in a different park, or listening to a new podcast—it's like a mini-spark for your brain.
- Track your progress with a "progress selfie" graph—focus on changes in your health numbers, like better sleep, not just what you see in the mirror.
- Keep 20% of your plan flexible to try new stuff, so you don't feel stuck. Use a meal-roulette wheel app to randomly pick Mediterranean dinners—like a spinach salad one night, fish the next—to keep eating exciting (*Dopamine and Novelty*).

17.2.2: STRESS SPIKES AND EMOTIONAL WHIPLASH

Stress can hit hard and mess with your emotions, but you've got ways to calm down fast.

- Remember the 90-second rule: feel the emotion, name it—like ***"I'm overwhelmed"***—and breathe deeply to let it fade without holding on.
- HRV biofeedback games turn calming down into a fun challenge, helping you get better at managing stress.
- Create a transition ritual—like taking deep breaths during your commute or pausing at your doorframe—to leave work stress behind before you walk into your home.
- If stress spikes your blood sugar, grab a protein snack, like a handful of nuts, instead of something sugary to keep it steady.

A corporate lawyer stopped a stress-eating binge by using 4-7-8 breathing—inhale for 4, hold for 7, exhale for 8—and sipping mint tea, finding calm in just a minute (*Stress and Emotional Regulation*).

17.2.3: MINDSET RE-FRAME TOOLBOX

Your thoughts can either hold you back or push you forward—let's reframe them to keep you going.

- Use a CBT "thought record" to catch negative thoughts—like ***"I'll never stick to this"***—challenge them with facts, and replace them with something better, like ***"I've got this one day at a time."***
- The WOOP method (*Wish, Outcome, Obstacle, Plan*) turns dreams into plans—wish to feel stronger, imagine the outcome, name the obstacle like lack of time, and plan to wake up 10 minutes earlier.
- Rewrite setbacks as part of your hero's journey, like the tough middle of a movie where you grow stronger.

• Set micro-goals—tiny 2-minute tasks like stretching—to get moving again. Write an email to your future self, scheduling it to arrive during a likely tough week, to remind you why you started (*CBT Thought Record*).

Of course it doesn't come easy! Nothing worth doing comes without effort. Remember what my friend, Secretary of State, General **Colin Powell** said:

"A dream doesn't become reality through magic; it takes sweat, determination and hard work."

Secretary of State, General Colin Powell and Tad Sisler

Source – Sisler Private Collection

SECTION THREE: COURSE-CORRECTION PROTOCOLS

When things don't go as planned, you can adjust your blueprint with some smart tweaks—like fine-tuning a car to keep it running smoothly. Let's set up some easy ways to course-correct.

17.3.1: DATA-DRIVEN ADJUSTMENT LOOPS

Your health numbers are like a map—they'll show you where to adjust if you're off track.

• Check your weekly dashboard: if something like your HRV or sleep drops way below your usual—two standard deviations, in tech terms—it's a red flag to make a change.

• Stick to the "one-variable" rule: tweak just one thing, like adding more protein, and check your numbers again after 14 days to see if it worked.

• Use a rolling 7-day average for your stats to avoid overreacting to a single bad day, like a rough night's sleep.

• An AI health-bot can suggest three fixes, ranked by how easy they are—pick the simplest one first, like walking more.

• Keep a "what worked" library—write down fixes that helped, like drinking more water, so you can use them again later (*Data-Driven Health Adjustments*).

17.3.2: MINIMAL-EFFECTIVE-DOSE FALLBACKS

When life gets crazy, you can still stay on track with quick, effective moves—like a start-up using shortcuts to keep growing.

• A 12-minute Tabata workout—short bursts of exercise like jumping jacks—gives 90% of the fitness boost of a 45-minute run, perfect for busy days.
• If you can't get to a cold plunge, dunk your face in cold water for 15 seconds to calm your nerves fast.
• Pack portable protein—like sardines or whey packets—for travel days when meals are tricky.
• Sleep-bank by adding an extra hour of rest for two nights before a red-eye flight to cushion the hit.

A sales rep hit her step goals by pacing airport halls between gates, showing how to sneak in movement anywhere (*Tabata and Fitness*).

17.3.3: ACCOUNTABILITY AND SUPPORT RE-BOOT

Your support crew can help you get back on track when you slip—like a team rallying to hit a deadline. Reactivate your accountability buddy with a quick 30-second voice memo challenge, like **"Let's both walk 5,000 steps today!"** Post a screenshot of your habit streak in your group chat every Sunday night to keep everyone cheering you on.

Host a quarterly "longevity potluck" to share wins and swap recipes—like a veggie-packed salad—to make healthy eating a group thing. If one area, like exercise, is really tough, hire a coach just for that to get the biggest bang for your buck (*Social Support and Adherence*).

Now that you can pivot under pressure, let's ride the wave of tomorrow's breakthroughs—without losing your footing.

CHAPTER EIGHTEEN
FUTURE-PROOFING YOUR LONGEVITY JOURNEY

What good is living to 150 if you're stuck with 2024 tech and 1990 mindsets? To thrive in the future, you need to keep up with the latest health breakthroughs and stay flexible enough to roll with the changes. In this chapter, we'll peek at what's coming in the next decade, build your adaptability muscles to handle anything, and lock in habits that keep you growing for a lifetime—making sure you're ready for whatever's next.

SECTION ONE: THE 2026-2036 INNOVATION HORIZON

The next 10 years are going to bring some mind-blowing health tech—like rocket fuel for your longevity journey. This chapter may be the most exciting one in the book, and we may not even be scratching the surface here of what's to come in the next handful of years. Let's check out what's on the horizon and how it can help you live better.

18.1.1: CELLULAR REBOOT THERAPIES FOR AGE REVERSAL

Want to get younger? Scientists are working on ways to "reboot" your cells, basically turning back the clock on aging at the cellular level.

- A method called partial Yamanaka reprogramming made mice live 24% longer by resetting their cells to a younger state—pretty amazing!

Starting in 2026, a human trial will test plasma exchange, where older people get plasma from younger donors to see if it perks up their health.

- A new pill called SSK-1, designed to clear out old, worn-out cells, is entering Phase II trials for knee arthritis and frailty, helping people move better.
- In monkeys, a gene-editing technique fixed a rare aging disease called progeria, hinting at future fixes for humans. In a first compassionate-use case, a patient who couldn't walk without a cane ditched it after this kind of therapy, showing what's possible (*Yamanaka Reprogramming*).

And this is all before we see what advancements in the use of stem cells and other new technology will have on the human condition. The future is wide open! My dear friend, former championship-winning *NFL* Quarterback, **Congressman** and **Secretary of Housing and Urban Development Jack Kemp** said:

> ***"There's no limit to what free men and free women in a free market with free enterprise can accomplish when people are free to follow their dream."***

Tad Sisler with Congressman Jack Kemp
Source – Sisler Private Collection

18.1.2: PERSONALISED OMICS AND DIGITAL TWINS

Soon, you'll be able to get a complete map of your health for a fraction of today's cost—like a personalized health GPS.

• By 2028, a whole-genome test will drop to $99, and a multi-omics bundle—checking your genes, gut, and more—will be under $300. This price is pretty amazing when you add up what it originally cost over a decade around the turn of the century to map the first human genome at around **$2.7 billion USD.**

• "Digital twins" of your organs will run overnight simulations to see how drugs might work for you, so you know what's safe before you try it.

• A gut-on-chip test lets you mail a sample to a lab and get back a heat-map showing how foods affect your gut—like which ones make you feel bloated.

• Wearable sweat-patches will track your nutrient levels in real time, letting you know if you need more vitamin D. Your data stays safe with cloud edge-computing, keeping it on your device instead of online (*Personalized Omics*).

18.1.3: SPACE-AGE WEARABLES AND SMART SURROUNDINGS

Wearables are getting a major upgrade—think sci-fi gadgets that keep you healthier without you even noticing.

• Graphene tattoo-sensors, as thin as a sticker, will stick to your skin and track stress hormones like cortisol and energy markers like lactate all day.

• Smart pajamas made with special fabric will send out far-infrared waves to help you sleep deeper, like a cozy hug from tech.

• Your ceiling lights will shift colors and brightness to match your body clock, helping you wake up or wind down naturally.

• AI air purifiers will sense when you're stressed—using your HRV—and clean out bad air particles to help you breathe easier.

• In an elderly home pilot, LIDAR floor sensors cut nighttime falls by alerting staff when someone got up, keeping people safer (*Smart Wearables and Health*).

SECTION TWO: BUILDING ADAPTABILITY MUSCLES

To keep up with all this new tech, you need to stay flexible—like stretching your mind and habits to handle change. Let's build those adaptability muscles so you're ready for the future.

18.2.1: LIFELONG LEARNING AND SKILL CYCLING

Learning new things keeps your brain young—it's like a gym workout for your mind. Your brain craves exercise just as your body does.

• Use the "learn-one-teach-one" trick: learn something new, then teach it to someone else to lock it in, which doubles how much you remember and boosts your brain's flexibility, as we saw in Chapter 9.

• Every three months, try a new skill sprint—like coding, playing an instrument, or dancing—to keep your memory center, the hippocampus, in top shape.

• AI tutors can give you quick lessons right when you need them, like a mini-teacher in your pocket. Combine skills—like learning about finance and genetics—to make yourself tougher against career changes. This builds cognitive reserve, which is like insurance against dementia, keeping your brain sharp for years (*Lifelong Learning and Brain Health*).

"If you put yourself in a position where you have to stretch outside your comfort zone, then you are forced to expand your consciousness."
– Les Brown

18.2.2: BUDGETING TIME AND MONEY FOR HEALTH TECH

Health and longevity shouldn't be just for the rich. Future-proofing means setting aside time and money for your health—like investing in a savings account for your body.

• Put 3% of your income into a "future-health fund" to cover things like new wearables or lab tests.

• Before buying gadgets, check if they're worth it—do they give you real insights, or are they just shiny toys with no payoff?

• Use your HSA or FSA accounts to pay for health stuff like blood tests or smartwatches if your plan allows it.

• Set aside 30 minutes each month on a Sunday for a "tech audit"—check what's working, like your step tracker, and what's not, like an app you never use.

I bought a smart TV and swapped my cable bill for a sauna subscription, saving money while boosting my health with weekly sessions (*Health Tech Budgeting*).

18.2.3: ETHICAL AND PRIVACY SAFEGUARDS

As you use more health tech, you've got to keep your data safe—like locking up your valuables.

- Always check "model cards" to see if the AI was trained on people like you, so it doesn't miss your needs.
- Pick devices that process data on the device itself, not in the cloud, and let you export your info to keep control.
- For genetic data, store it on an encrypted offline drive with two-factor authentication, like a super-secure vault.
- Make a "digital advance directive" to decide what happens to your data after you're gone—think of it as a will for your info.
- If you join citizen-science projects, make sure they're open about how they use your data, so you're not left in the dark (*Data Privacy in Health Tech*).

It's as important today to keep ethics and morality on the front burner when addressing new technology as it was in the 1980s when my friend, **President George H. W. Bush**, said:

"America is never wholly herself unless she is engaged in high moral principle. We as a people have such a purpose today. It is to make kinder the face of the nation and gentler the face of the world."

President George H.W. Bush and Barbara Bush with Tad Sisler

Source – Sisler Private Collection

SECTION THREE: LOCKING IN THE "FOREVER SWITCH"

Remember the "forever switch" in our introduction? To make your longevity journey last, you need habits that stick and a vision to keep you inspired—like flipping that switch that stays on for life. Let's make sure your plan is set for the long haul.

18.3.1: QUARTERLY INNOVATION AUDIT

Every three months, do a quick check on your health tech—like cleaning out your closet to make room for new stuff.

- Use a 4-step checklist: assess what tech you're using, triage what's working or not, pilot a new tool for a month, then integrate the ones that help.
- Score new gadgets on safety, evidence, cost, and how fun they are—don't keep something that's a hassle.
- Sunset old tools that don't work anymore, like an outdated app, to declutter your data and devices.
- Keep a "Do-not-try" list for hyped-up stuff that's not worth it, like a pricey gadget with no proof it helps.
- Make a vision board of new things to try, like a smart air purifier, to keep you excited (*Innovation Audit for Health*).

18.3.2: LEGACY AND MENTORSHIP

Sharing what you've learned can give your life more meaning—like planting a tree for the next generation to enjoy.

- Teach your 90-day blueprint to a younger person, like a teen or grandchild—it'll boost your purpose hormones, like oxytocin, making you feel connected and happy.
- Record a podcast of your family's health stories, which can pass down resilience habits that even affect your genes, as we saw in Chapter 13.
- Volunteer in a community health project, like a walking group, to give back and get feedback on what works.
- Set up a "health endowment"—a small fund of time or money—that grows with you, like donating to a local gym.

My 83-year-old friend **Mike**, a former local basketball coach, was still mentoring teens on the court, and his HRV beat his peers by 15 milliseconds, showing how giving back keeps you healthy (*Mentorship and Purpose*).

18.3.3: FUTURE-SELF VISUALIZATION AND COMMITMENT

Let's make sure your future self is cheering you on—like writing a letter to the you of tomorrow.

Do a 5-minute guided visualization: picture yourself at 120, waking up full of energy, maybe doing yoga or laughing with family—what does that morning look like? Write a postcard from that future you, describing how great you feel, and mail it to yourself to arrive a year from now as a reminder. Pick a big, dreamy goal—like running a marathon in space tourism—to keep you inspired, no matter how wild it sounds. Sign a contract with your accountability partner to review your goals every equinox, like a seasonal check-in to stay on track (*Future-Self Visualization*).

All systems go. In the final pages, we'll stitch every thread—mind, body, tech, and community—into a lifetime loop of limitless living. But first, let's look at other unbelievable aspects of the mind that go way beyond the norm.

CHAPTER NINETEEN

UNEXPLAINED BRAIN PHENOMENA: UNLOCKING THE MIND'S MYSTERIES

Now we have our action plan, and we're working on future-proofing our longevity journey. But there is so much more to us than we realize. Your brain is like a mysterious universe—it can do things that seem almost magical, from sensing the future to blending your senses in wild ways. These unexplained phenomena might sound like sci-fi, but they're real experiences that show just how incredible your mind is. And... unique to the human experience, these "powers" differentiate us from quantum computers and robots.

In this chapter, we'll explore some of the strangest brain mysteries, from seeing colors in music to feeling like you've lived this moment already, and see how they might even help you live a healthier, longer life.

Here's a caveat: This chapter may not be for everyone, so I saved it for last. If this does not resonate with you, that's ok. Some people have actually experienced some or all of the phenomena I'm discussing here, and it's amazing and sometimes unexplainable. If you've ever experienced any of these phenomena, consider forming your own action plan to develop these skills, which seem to be stronger in some people than in others.

SECTION ONE: SENSING BEYOND THE EVERYDAY

Your brain sometimes picks up on things that feel beyond normal—like a sixth sense or a glimpse into the future. Let's dive into some of these mind-bending experiences.

19.1.1: PSYCHIC ACTIVITY AND PREMONITIONS

When I was twenty, I was managing a restaurant. I had worked there for a couple years, and I knew the owner was happy with my work.

Even still, he hired a new General Manager, and this man wanted to bring in his own staff. I was unaware of this, but I had a vivid dream the night before of walking into an office I had never seen at the restaurant and sitting across from him as he let me go. The next morning, I went into the restaurant, and a waiter told me that the new GM was looking for me. He had renovated an old storeroom into his office on my days off. As I walked into the new office, my hair stood up on end. It was exactly the room I had dreamed of. Even the picture on the wall was the same as in my dream. The words he spoke were exactly what I heard in my dream, word for word. I had experienced a real premonition.

So, as far-fetched as it may seem, I know it's real from my own experience.

Have you ever had a gut feeling about something that turned out to be true—like knowing a friend was in trouble before they called? That's what some call psychic activity, like telepathy or clairvoyance, where you seem to sense things without using your usual senses. Premonitions are similar—a strange feeling that something's about to happen, like dreaming of an event before it unfolds. Scientists are still figuring this out, but some studies, like the Princeton Global Consciousness Project we mentioned in Chapter 3, suggest our minds might connect in ways we don't fully understand, especially during big emotional moments. While it's not something you may be able to practice, being open to these intuitions might help you trust your gut, which can lower stress and guide better choices (*Princeton Global Consciousness Project*).

You've heard about "women's intuition." I honestly believe in it, because I've seen it in action with my mother, sisters, and wife. I might not always like to admit it, but more often than not, they're right when they feel something strongly. Don't bet against it.

19.1.2: OUT-OF-BODY EXPERIENCES (OBEs)

Ever felt like you were floating outside your body, looking down on yourself? That's an out-of-body experience, or OBE, where you feel like you're seeing the world from somewhere else. It often happens during intense moments, like trauma or deep meditation, and some people report it during near-death experiences. Researchers think it might come from a mix-up in your brain's sensory areas, like the temporoparietal junction, which handles where you feel your body is. While OBEs can be spooky, they show how flexible your brain is—some people even use them in meditation to feel more connected to the world, which can reduce stress and boost well-being (*OBEs and Brain Activity*).

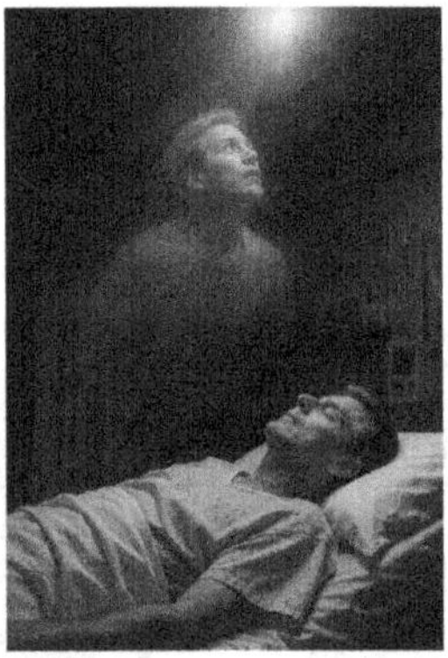

19.1.3: MANIFESTING HIGHER CONCIOUSNESS THROUGH THE UNIVERSE

Ever heard of putting your thoughts out into the universe and watching them come true—like wishing for a new job and landing it? That's the idea of higher consciousness, where your mind connects to something bigger, almost like tuning into a cosmic radio station. People who practice manifesting, like vision board fans, say focusing on a goal—like picturing yourself healthy and active at 90—can make it more likely to happen because your brain gets laser-focused, boosting motivation and spotting opportunities, kind of like how optimism rewires your brain for better outcomes (Chapter 14). While science can't yet prove the "universe" listens, **Ernest Holmes** book *The Science of Mind* explains well how this concept works—and studies show that strong belief can lower stress and spark feel-good chemicals like dopamine, helping your body stay energized and resilient, which we know adds years to your life (*Belief and Dopamine*).

"Life is a mirror and will reflect back to the thinker what he thinks into it." – Ernest Holmes

Ernest Holmes

Credit – Chat GPT Image Generated

19.1.4: WHITE LIGHT PROTECTION: A SHIELD OF ENERGY

Surrounding someone with a white light of protection is a spiritual practice where you imagine a glowing, pure light wrapping around you or a loved one—like a shield of good vibes to keep negativity away. It's often used in meditation or prayer, like we talked about in Chapter 13, and people say it makes them feel safe and calm, whether they're facing a tough day or sending love to someone far away. Your brain might be tapping into a higher consciousness here, but it's also doing something pretty cool on its own—visualizing that light can activate your brain's calming centers, lowering stress hormones like cortisol, much like the placebo effect we mentioned in Section 19.3.1, giving your body a little boost of peace and strength (*Visualization and Stress*).

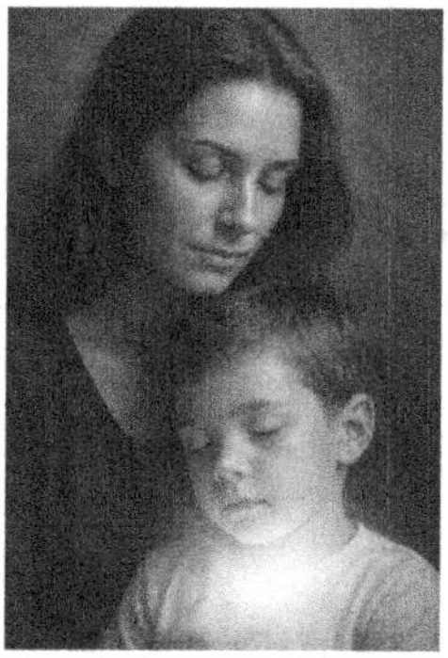

SECTION TWO: THE BRAIN'S CREATIVE AND SENSORY QUIRKS

Your brain doesn't always play by the rules—it can mix things up in ways that feel like magic. Let's explore some of these quirky brain tricks and what they mean for you.

19.2.1: SYNESTHESIA: BLENDING YOUR SENSES

Imagine hearing a song and seeing bright colors dance in front of you—that's synesthesia, where your senses get all mixed up. Some people "see" sounds, "taste" colors, or even "feel" numbers as textures. It's like your brain's wires cross in a cool way, connecting senses that usually stay separate. About 4% of people have this, and it's often linked to extra connections in the brain's sensory areas. Synesthetes, as they're called, often say it makes life more vivid—like artist **Pharrell Williams**, who sees colors in music and uses it to create his hits.

For you, this shows how unique your brain can be, and even if you don't have synesthesia, trying new sensory experiences—like listening to music while painting—can spark creativity and reduce stress (*Synesthesia and Brain Wiring*). It's kind of like a new way of multitasking!

"It's not possible to experience constant euphoria, but if you're grateful, you can find happiness in everything." – Pharrell Williams

Pharrell Williams

Credit – Wikimedia Commons

19.2.2: LUCID DREAMING AND HYPNAGOGIC STATES

Ever been in a dream and suddenly realized you're dreaming? That's lucid dreaming, where you're aware and can even steer the dream—like flying or exploring a fantasy world. It often happens during REM sleep, and some people train for it by keeping a dream journal or checking if they're awake during the day. Then there are hypnagogic and hypnopompic states—the dreamy moments as you fall asleep or wake up—where you might see vivid images or get creative ideas, like a melody popping into your head. Famous inventor Thomas Edison used these states by napping with a ball in his hand, waking as it dropped to capture ideas. These states can boost creativity and even help you relax, giving your brain a playful break that might improve your mood (*Lucid Dreaming and Creativity*).

19.2.3: Déjà Vu, FALSE MEMORIES, AND SAVANT SYNDROME

When my twins were born, my son was put into an incubator for a couple days so his lungs could fully develop. At some point on that wonderful day, a nurse brought our daughter into the room and placed her in my arms. She looked up at me and we immediately recognized each other! I don't know from where, but I absolutely knew this person, this lovely little soul, and my first thought was, ***"we're together again!"***

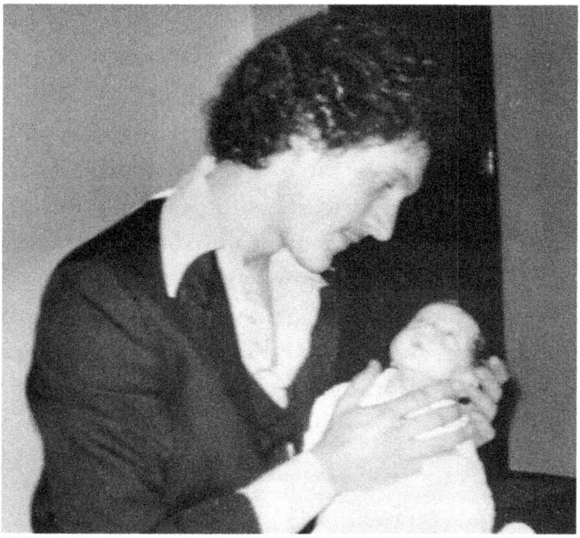

Tad Sisler with Daughter Rachel at her Birth

Source – Sisler Private Collection

• Déjà vu is that eerie feeling when something new feels oddly familiar—like you've lived it before. Scientists think it might be a memory glitch in your temporal lobe, making a new moment feel old.

• False memories are even wilder—your brain can "remember" things that never happened, like recalling a party you didn't attend, because memory is more like a story that changes over time.

• Then there's savant syndrome, where someone with a developmental difference, like autism, has an amazing skill—like doing huge math problems in their head or playing music perfectly after hearing it once.

These quirks show how mysterious your brain is, and while you might not control them, they remind you to be curious about your mind, which can inspire learning and growth (*Déjà Vu and Memory*).

SECTION THREE: MIND-BODY MYSTERIES WITH HEALTH BENEFITS

Some brain phenomena don't just sound cool—they might actually help you live better and longer. Let's see how these mysteries can work for you.

19.3.1: PLACEBO EFFECT: THE POWER OF BELIEF

The placebo effect is like a superpower—your belief in a treatment can make you feel better, even if it's just a sugar pill. If you think you're getting medicine, your brain can release feel-good chemicals like dopamine, easing pain or boosting your mood. Studies show this can help with things like headaches or even depression, sometimes as much as real drugs. It's why believing in your health plan—like the 90-day blueprint from Chapter 16—can make it work better. You can tap into this by staying positive about your habits, like trusting that your daily walk is healing you (*Placebo Effect and Health*).

19.3.2: SPLIT-BRAIN PHENOMENA AND BEYOND

Imagine your brain as two halves that usually work together—but in split-brain patients, they're separated because the bridge between them, called the corpus callosum, is cut to help with seizures. This leads to wild effects, like one hand buttoning a shirt while the other unbuttons it, showing how each half can act on its own. It's rare, but it teaches us how flexible our brains are, which ties into the neuroplasticity we explored in Chapter 9. While you probably don't have a split brain, knowing it can adapt in such big ways reminds you that your mind can handle challenges—like learning a new skill or bouncing back from stress (*Split-Brain Phenomena*).

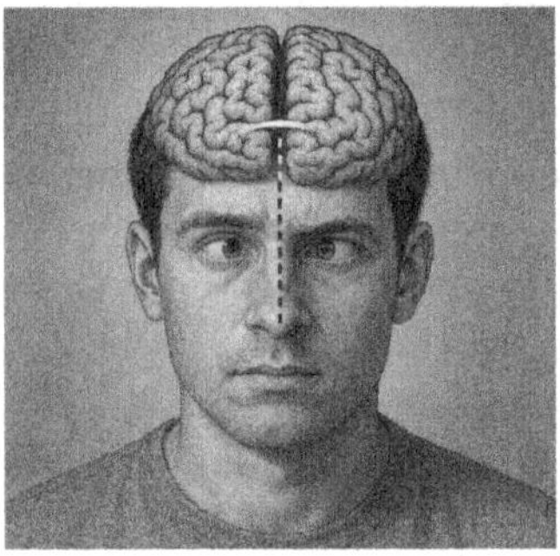

19.3.3: WHAT THIS MEANS FOR YOUR LONGEVITY

These brain mysteries aren't just party tricks—they show how powerful your mind is, and that can help you live longer.

• Psychic activity and premonitions might encourage you to trust your instincts, reducing stress when you listen to your gut.

• Synesthesia and lucid dreaming can spark creativity, which lowers anxiety and keeps your brain young.

• Déjà vu and false memories remind you to question your assumptions, keeping your thinking sharp. Or, maybe they're a door into a past beyond this lifetime we can't remember.

• The placebo effect proves your beliefs can heal you, so stay positive about your health journey.

• And split-brain phenomena show your brain's ability to adapt, which can inspire you to keep learning and growing.

19.3.4: AUTHENTICATED MIRACLES: FAITH AND THE MIND

Miracles might sound like something out of a storybook, but there are real, verified cases that show how faith can do wild things to our brains and bodies—like a little nudge from the universe.

Take the 1980s case of a young girl in Medjugorje, Bosnia, who was blind from birth but suddenly gained her sight during a religious event, a miracle investigated and confirmed by medical experts who couldn't explain it (*Divine Mysteries: Queen of Peace*).

When your mind latches onto faith—like believing in a higher power during prayer, as we saw in Section 13.2.1—it can light up parts of your brain that handle hope and calm, like the anterior cingulate, lowering stress hormones like cortisol. That deep belief can even trigger your subconscious to release healing chemicals, kind of like the placebo effect we just talked about, helping your body fight off illness or pain and pushing you to live with more purpose, which we know adds years to your life.

"To one who has faith, no explanation is necessary. To one without faith, no explanation is possible." – Thomas Aquinas

Thomas Aquinas

Credit – Chat GPT Image Generated

19.3.5: NEAR-DEATH EXPERIENCES: A GLIMPSE BEYOND

Near-death experiences, or NDEs, are when someone comes extremely close to dying—like their heart stops—and they see things that blow their mind, like bright lights or loved ones who've passed.

In 2014, a man named **Jim** had a heart attack and flatlined for 10 minutes; he later shared how he felt total peace, saw a glowing tunnel, and even watched doctors working on him from above, details doctors confirmed he couldn't have known.

Science says these might come from your brain going haywire under stress—like a burst of activity in your temporal lobe—but your subconscious can process it as a spiritual wake-up call, making you less scared of death and more focused on living with joy, as folks often report after NDEs. Although I look at these experiences more profoundly as proof that our souls continue to exist beyond this life, at the very least that spiritual wake-up call can lower anxiety, boost your mood, and even improve your health, giving you a fresh perspective to tackle life's challenges with a calmer, stronger body and mind.

19.3.6: BIOPHOTONS: THE BODY'S HEALING LIGHT

Your body is doing something pretty amazing right now—it's creating tiny bursts of light called biophotons, which are like invisible signals helping your cells talk to each other and heal. These ultra-weak light emissions come from your metabolic processes, especially when reactive oxygen species (ROS)—think of them as little energy sparks—interact with your cells, like in your mitochondria, the powerhouses we discussed in Chapter 11. Scientists, like biophysicist **Fritz-Albert Popp**, found that healthy cells give off more stable biophotons than sick ones, suggesting this light might coordinate healing by keeping your cells in sync, almost like a secret language your body uses to fix itself.

This light isn't just floating around—it's like food for your cells, with some researchers saying you "eat" the light stored in the food you consume, like sucking energy from a battery to recharge your body's healing powers. By eating nutrient-rich foods—like the fruits and veggies we talked about in Chapter 8—you might be boosting your body's light energy, which could help lower inflammation and speed up recovery, giving you another tool to live longer and healthier, tying right into our focus on natural healing throughout the book.

Every concept in this book is aimed at helping you to know yourself on a higher level. If you can achieve this at any age, you should feel more fulfillment and confidence along with health and longevity benefits that will naturally come.

This reminds me of what my good friend, legendary *Academy-Award-nominated* screen actor **Elliott Gould** said:

"My problem was I let myself become known before I knew myself."

Tad Sisler with Elliott Gould

Source – Sisler Private Collection

Don't worry. There's always time if you begin the process now. Now, let's wrap it all up with a recap.

CONCLUSION

THE LOOP OF LIMITLESS LIVING

Remember **Angela,** the 128-pound mom who lifted a two-ton SUV off her son? In that nine-second surge, she proved the thesis of this book: the "forever switch" already lives inside you. Everything you've read—from mitochondrial tune-ups to quantum dashboards—has been about learning to flip that switch on purpose, day after day. You've got the tools to reprogram your mind, body, and genes for a longer, more vibrant life, and now it's time to put it all together into a loop that keeps you thriving—no matter what comes your way.

SECTION ONE: WEAVING THE JOURNEY

Let's pull together the big lessons from your journey—like weaving a strong rope from all the threads of this book. Mind, body, tech, and tribe all work together to help you live your best life.

C.I.I: MIND - SOFTWARE OVER HARDWARE

Your mind is like the software running your body's hardware—it can make all the difference in how long and well you live. Focused thoughts can spike adrenaline for a quick energy boost, like Angela did, and rewire your brain's connections, as we saw with neuroplasticity in Chapter 9. Tools like CBT, mindfulness, and flow-state drills from Chapter 14 act like a self-updating operating system, keeping your brain sharp and calm.

Optimism isn't just a feel-good vibe—a meta-analysis showed it can add 7.2 years to your life by lengthening your telomeres, those protective caps on your DNA (Chapter 14). The story loops we talked about—turning setbacks into growth—give your brain a neurochemical boost, like fuel for resilience. A reader emailed us saying she used the breath-count trick from Chapter 14 to avoid a panic attack on a turbulent flight, proving how these mind hacks can work in real life (*Optimism and Longevity*).

C.I.2: BODY - BIOTECH YOU ALREADY OWN

Your body is a powerhouse of biotech you were born with—no lab needed!

• Sleep, as we learned in Chapter 11, repairs 3 million cells every minute—it's the cheapest, most powerful drug on earth, rebuilding you while you dream (and maybe even lucid dreaming, as we explored in Chapter 19).

• Hormesis—using heat, cold, and fasting from Chapter 12—teaches your cells to be antifragile, getting stronger under stress, just like Angela's adrenaline surge.

• Your gut, that "rainforest" we dove into in Chapter 8, shapes your immunity, mood, and weight by working with your brain, sometimes even mixing senses in wild ways like synesthesia (Chapter 19).

• Tracking metrics like HRV, CGM, and grip strength turns your gut feelings into clear dashboards, giving you data to act on (Chapter 16).

Angela's strength came from adrenaline and her brain overriding her limits—now you know how to train that safely with the tools in this book (*Sleep and Cellular Repair*).

C.I.3: TECH AND TRIBE - EXPONENTIAL TAILWINDS

Tech and your community are like tailwinds pushing your longevity journey forward—making everything easier and faster.

• AI wearables from Chapter 15 can spot trouble before you feel it, like catching a heart rhythm issue early, using a predict-prevent loop to keep you healthy.

• Quantum tech, also in Chapter 15, screens new anti-aging drugs while you sleep, bringing cutting-edge solutions closer every day.

• Family dinners, as we saw in Chapter 13, spike oxytocin—the bonding hormone—lowering your blood pressure and inflammation markers like CRP, all while passing down resilience through shared stories.

• Community challenges, like cold-plunge clubs, boost your stick-to-it rate by 1.4 times, making health a team sport (Chapter 13).

• And keeping your moral compass and data privacy in check, as we covered in Chapters 15 and 18, ensures tech stays human-centered—not a takeover (*Community and Adherence*).

SECTION TWO: YOUR THREE-RING OPERATING SYSTEM

Let's revisit the 90-day blueprint from Chapter 16—like a simple operating system to keep your longevity loop running smoothly. It's got three rings to balance your mind, body, and recovery, so you're always moving forward.

C.2.1: DAILY CORE (15 min × 3)

Your daily core is the heartbeat of your plan—small, quick habits that add up to big wins over time.

- For your mind, take 60 seconds to do box breathing—inhale for 4, hold for 4, exhale for 4, hold for 4—then jot down something you're grateful for, like a good cup of coffee, to boost your mood (Chapter 14).
- For your body, do a movement snack: try door-frame push-ups or sun salutes for a quick stretch and strength boost, keeping you active without a gym (Chapter 16).
- For recovery, take a 2-minute cold rinse in the shower—it's like a mini hormesis hit—and start a digital sunset an hour before bed, turning off screens to help you sleep better (Chapter 12).

Stack these onto habits you already have, like breathing while your coffee brews or stretching at lunch, to make them stick. Aim for 80% consistency—if you hit most days, you'll see those gains compound over time (*Box Breathing and Stress*).

C.2.2: WEEKLY UP-SHIFT

Each week, add some bigger moves to level up—like giving your operating system a little upgrade.

- Do two Zone-2 cardio sessions, like a brisk walk where you can still chat, two strength workouts to keep your muscles strong, and one sauna or cold plunge to build resilience, as we learned in Chapter 12.
- Spend 120 minutes in green space—like a hike or gardening—to reset your stress hormone cortisol and soak in nature's benefits (Chapter 10).

• On Sundays, meal-prep a Mediterranean-plus menu—think veggies, olive oil, and fish—to set yourself up for healthy eating all week (Chapter 16).
• Plan a dinner with family or friends to boost oxytocin and keep you accountable, sharing stories that might even spark a déjà vu moment (Chapters 13, 19).
• Review your dashboard—color-code your wins and gaps, like green for good sleep and yellow if your steps are low—to see where you're at (*Nature and Cortisol*).

C.2.3: QUARTERLY RE-CODE

Every three months, tweak your plan to keep it fresh—like updating your phone to the latest version.
• Test your microbiome and epigenetic clock to see how your gut and biological age are doing, then adjust your fiber intake or fasting schedule, like adding more kimchi or a 48-hour fasting-mimicking reboot (Chapters 8, 11, 16).
• Spend a month swapping skills—learn a new language, instrument, or sport—to keep your brain growing, maybe even tapping into hypnagogic creativity for inspiration (Chapters 9, 19).
• Do a "digital detox retreat"—take a weekend off screens—to reset your dopamine levels, helping you feel refreshed (Chapter 18).
• Check your North-Star Goal from Chapter 16: does your calendar match what matters most to you, like family or health? This keeps you aligned with your big dreams (*Microbiome Testing*).

SECTION THREE: LAUNCH SEQUENCE - 90 DAYS TO PROOF

You're ready to launch your limitless living loop—let's get started with a 90-day sprint to prove it works. Think of this as your countdown to liftoff, with everything you need to take flight.

C.3.1: ACTIVATE THE PLANNER (Appendix A)

It's time to put your plan into action—like flipping the switch to start your rocket.
• Turn to Appendix A and transfer your baseline labs and metrics—like your HRV or sleep score—into the page 1 grid to know where you're starting.
• Circle one keystone habit per area—mind, body, recovery—like gratitude for mind or push-ups for body, to focus your efforts.
• Set a start date within 72 hours—momentum loves speed, so don't wait! Invite an accountability partner, like a friend or family member, and share your dashboard link so they can cheer you on.
• Print the tracker and stick it on your fridge or set it as your phone's lock screen to keep it front and center (*Action Planning and Momentum*).

C.3.2: TRACK, TWEAK, TRIUMPH!

Now, let's make sure you stay on course—like a pilot checking the gauges mid-flight.

- Log your daily streaks in your tracker—did you do your breathing or movement?—and celebrate micro-wins, like hitting your step goal, with a dopamine-anchor ritual, maybe a quick fist pump to feel good (Chapter 17).
- If a metric, like your HRV, drops way off your baseline—two standard deviations in tech terms—it's a red flag; use the pivot tricks from Chapter 17, like a quick cold rinse, to get back on track.
- Every 30 days, do a mini-retrospective: what's working (keep), what needs a tweak (adjust), and what's not (drop), keeping your plan fresh.
- Snap photos of your progress graphs—like better sleep scores—to reinforce your new identity: ***"I'm a longevity engineer"*** (*Tracking and Adjustment*).

C.3.3: PAY IT FORWARD AND FUTURE PROOF

Your journey isn't just for you—it's a legacy to share, like passing down a family recipe.

- Teach one element of your blueprint to a friend or child, like how to do box breathing—it locks in your learning and spreads the health vibes, boosting your purpose hormones (Chapter 13).
- Record a 60-second "future-self" video—imagine yourself at 90, full of energy, maybe even tapping into a placebo effect belief to feel stronger (Chapter 19)—and set a phone reminder to watch it a year from now.
- Put 3% of your income into a "health-future fund" to try new tech, like the smart pajamas from Chapter 18, keeping you ready for what's next.
- Do a quarterly innovation audit—score new gadgets on safety, payoff, and joy, like we talked about in Chapter 18, to stay ahead without clutter.

Your legacy mindset means the habits you model—like staying curious, maybe even exploring psychic intuitions (Chapter 19)—can become someone else's forever switch (*Legacy and Purpose*).

FINAL CALL-TO-ACTION

Thank you so much for reading this book! I've also learned volumes while writing my **Health and Longevity Mastery Series**, and I hope you benefit from this to the point where we can talk about it together fifty years from now!

Start by turning the page to Appendix A. In ten minutes, you'll have a personalized 90-day roadmap. In 90 days, you'll have data-proof that your mind, body, and genes really can cooperate to add vibrant years. Share your Day-0 metrics with a partner tonight and start in the morning! The **future you** is waiting—don't keep him or her on hold. And don't lose the faith! If you get off-track, remember, every day you can wake up and reinvent yourself!

APPENDICES

These appendices are your toolkit to make the 90-day longevity blueprint from Chapter 16 come alive. They're designed for you to create your own csv or excel documents, scan and print them out, or use them digitally. You'll find everything you need to track your progress, pick the right supplements, take action, find resources, and understand key terms—all in plain English, so it's easy to jump in and get started.

Appendix A – Baseline Lab & Metric Sheet

This sheet is your starting point—like a health map to see where you're at and where you're going. Record your numbers before you start (Day-0), then again at Day-30 and Day-90 to track your progress. Create color codes to see how you're doing: Green means you're on target, Yellow means keep an eye on it, and Red means it's time to make a change (check Chapter 17 for pivot ideas).

Domain	Metric	Target Range	Your Day-0	Day-30	Day-90
Cardio-Metabolic	Resting HR (bpm)	50-70			
	HRV (RMSSD, ms)	>50 (40+)			
	Blood Pressure (mmHg)	<120/80			
	Fasting Glucose (mg/dL)	70-90			
	ApoB (mg/dL)	<80			
	Triglycerides (mg/dL)	<100			

Inflamm-Immune	hs-CRP (mg/L)	<1.0
	Homocysteine (µmol/L)	<10
	Vitamin D (ng/mL)	40-60
Hormones	Morning Cortisol (µg/dL)	6-18
	Free T3 (pg/mL)	2.5-4.2
	Total Testosterone / Estradiol	Age-adjusted
Body Comp	Waist-to-Height Ratio	<0.5
	Lean Mass (DEXA %)	↑ trend
Longevity Clocks	DNAmAge (Horvath/GrimAge)	≤ chrono-age
	Telomere Length (kb)	≥7.0

How to Use: Fill in your Day-0 numbers before starting the 90-Day Blueprint. Repeat at Day-30 and Day-90. Green = on target, Yellow = watch, Red = intervene (use Chapter 17 pivot protocols to adjust).

Appendix B – Supplement & Peptide Cheat Cards

Think of this as your quick guide to supplements—like a cheat sheet for boosting your health. They're split into tiers based on your goals, with doses, benefits, and cautions in plain English. Always check with your doctor before starting, especially for peptides.

Tier	Compound	Dose / Timing	Primary Benefit	Evidence Note / Caution
Core Daily Stack	Magnesium Threonate	2 g, bedtime	Deep sleep, brain connections	Stomach upset if over 2 g
	Omega-3 DHA/EPA	1–2 g, with fat meal	Heart and brain health	Check TG-EPA ratio for balance
	Vitamin D3 + K2	2,000 IU + 100 µg	Immunity, calcium movement	Test vitamin D levels every 6 mo
	Creatine Monohydrate	5 g, morning	Strength, brain power	Drink plenty of water

Advanced "Mito" Stack	CoQ10 (Ubiquinol)	200 mg, breakfast	Boosts cell energy production	Caution if on blood thinners
	PQQ	10-20 mg, morning	Grows new energy factories	Works best with CoQ10
	Urolithin A	500 mg, empty stomach	Clears out old cell parts	Needs a healthy gut to work
Adaptogen / Nootropic	L-Theanine + Caffeine	200 mg : 100 mg, morning	Calm focus without jitters	Don't take after 2 p.m.
	Lion's Mane	1,000 mg twice daily	Boosts brain growth, memory	Avoid if allergic to mushrooms
	Rhodiola Rosea	200 mg before workout	Handles stress, boosts stamina	Don't take at bedtime
Senolytic Weekend (qtrly)	Fisetin	1,500 mg x2 days	Clears "zombie" cells	Take with a fat meal
	Quercetin + Dasatinib*	1,000 mg + 100 mg*	Experimental combo for old cells	*Prescription only, check with MD
Peptides (Rx / MD supervision)	BPC-157	250 μg injection daily x 30 days	Repairs gut, tendons, muscles	Cycle use, consult MD
	MOTS-C	10 mg injection, 2×/week	Boosts metabolism	Limited human data, check with MD
	FOXO4-DRI	Trial-phase	Kills off old, stuck cells	Monitor liver with MD

Appendix C – One-Page Action Worksheets

These worksheets boil down each chapter into a single page of "Do-Now" actions—like a cheat code for every lesson in the book. You cancreate your own csv or excel documents, scan and print them, tear them out, or load them into apps like Notion or Obsidian. Each follows the same grid: quick wins to start fast, lifelong habits to keep going, and metrics to track your progress.

Sample: Chapter 5 – Sleep Your Way to 120

Quick Wins (72-hr Jump-Start)	Lifelong Integration	Metrics to Watch
• Dim house lights to 50 lux or less after 8 p.m.	• ChiliPad at 18 °C for cool sleep.	Oura deep-sleep %, HRV, morning fasting insulin.
• Finish last meal 3 hours before bed.	• Dawn-simulator alarm to wake naturally.	
• Do 4-7-8 breath cycle (inhale 4, hold 7, exhale 8) ×4 sets at lights-out.	• Wear blue-light blocker glasses every night.	

(Worksheets for all 19 chapters follow this grid—use one page per chapter to keep your plan handy.)

Appendix D – Resource Directory

This directory is your go-to list for trusted places to get tests, gear, and support—like a map to the best tools for your longevity journey. Whether you're looking for labs, apps, or communities, we've got you covered.

Labs & Clinics

Name	Specialty	Region	Notes
Buck Institute Executive Panel	Multi-omic age clocks	USA (CA)	2-day intensive program
Potocsnak Longevity Clinic	Hormones + peptides	USA (IL)	MD/DO team for personalized care
TruDiagnostic	At-home DNAmAge	Global mail-in	GrimAge, PhenoAge tests
Nordic Gut Lab	Shotgun microbiome	EU/US kits	AI diet recommendations

Apps & Wearables

Tool	Function	Platform
Zero	Fasting tracker	iOS / Android
Ōura Ring Gen 3	Sleep + HRV tracking	Ring
Levels / Nutrisense	CGM insights	iOS / Web

Apollo Neuro	Vagal-stim touchband	Strap
NatureDose	Green-time monitor	iOS

Podcasts

• **Huberman Lab** – Neuroscience-backed protocols to boost your health and brain.
• **The Drive (Peter Attia)** – Deep-dive longevity science with expert interviews.
• **Lifespan (David Sinclair)** – Updates on aging reversal and cutting-edge research.

Communities

• **r/Biohackers (Reddit)** – Crowd-sourced feedback on health protocols.
• **Blue Zones Online** – Lifestyle challenges and recipes inspired by centenarians.
• **Faith-Centered Health Coalition** – Spiritual wellness groups for holistic health.

Appendix E – Glossary of 100 Longevity Terms

This glossary is your quick guide to key terms in THE UNLIMITED POWER OF YOUR MIND AND BODY, making complex ideas about longevity, health, and mind-body connections crystal clear. We've expanded from 40 to 100 terms, covering psychology, neuroscience, genetics, nutrition, exercise, and holistic health, with plain English definitions to help you understand and apply the book's concepts. Terms reflect the latest 2025 research for accuracy and relevance.

- **Adiponectin**: A hormone from fat cells that helps burn fat and keep blood sugar steady.
- **Aerobic Exercise**: Activities like running or cycling that boost heart rate and improve heart health.
- **AMPK**: A "fuel gauge" in cells that turns on fat-burning and cleanup when energy is low.
- **Antioxidants**: Molecules in fruits and veggies that protect cells from damage by free radicals.
- **Apoptosis**: Programmed cell death to remove damaged cells, keeping your body healthy.
- **Autophagy**: Your cells' recycling system, breaking down old parts to build new, healthy ones.
- **Biological Age**: How old your body seems based on health markers, not your actual age.
- **Biohacking**: Using science and tech to optimize your body's performance and health.

- **Blood Pressure:** The force of blood against artery walls; high levels can signal heart risks.
- **Blue Zones:** Five regions where people often live past 100, offering lifestyle tips for longevity.
- **BMR (Basal Metabolic Rate):** Calories your body needs at rest to stay alive.
- **Body Composition:** The ratio of fat, muscle, and bone in your body.
- **BPC-157:** A peptide that speeds healing in your gut, tendons, and muscles.
- **Brain-Derived Neurotrophic Factor (BDNF):** A protein that helps brain cells grow and survive, aiding memory.
- **Brown Fat:** Special fat that burns calories to create heat, often activated by cold.
- **Caloric Restriction:** Eating fewer calories without malnutrition to potentially extend lifespan.
- **Calorie Restriction Mimetics:** Compounds mimicking caloric restriction's benefits without less food.
- **CGM (Continuous Glucose Monitor):** A wearable tracking blood sugar every few minutes.
- **Chronotype:** Your natural sleep-wake preference, like being a morning lark or night owl.
- **Circadian Rhythm:** Your body's internal clock regulating sleep and other processes.
- **Cognitive Behavioral Therapy (CBT):** Therapy to change negative thought patterns for better mental health.
- **CoQ10:** An antioxidant helping cell energy factories produce power.
- **Cortisol:** Your stress hormone, helpful in short bursts but harmful when always high.
- **CRISPR:** Gene-editing tool that cuts and replaces DNA to fix health issues.
- **Cryotherapy:** Using cold temperatures to reduce inflammation and aid recovery.
- **Curcumin:** Turmeric's active compound with anti-inflammatory and antioxidant effects.
- **Default Mode Network (DMN):** Brain areas active during rest, linked to daydreaming and self-reflection.
- **DNAmAge:** A test showing your biological age based on DNA patterns.
- **Dopamine:** Brain chemical that boosts motivation and feelings of reward.
- **Emotional Intelligence:** Recognizing and managing your emotions and others' for better relationships.
- **Endorphins:** Hormones that reduce pain and boost pleasure, often from exercise or laughter.
- **Epigenetics:** Chemical switches turning genes on or off without changing DNA.
- **Exosome Therapy:** Using cell-derived vesicles to promote healing and regeneration.
- **Fasting-Mimicking Diet (FMD):** A 5-day, low-calorie plan mimicking fasting's benefits.
- **Fisetin:** A plant compound that may clear out harmful senescent cells.
- **Flow State:** Being "in the zone," fully focused and productive.
- **Forest Bathing (Shinrin-yoku):** Spending time in nature to reduce stress and boost health.
- **Free Radicals:** Unstable molecules that damage cells and speed aging.

- **Gene Expression:** How genes are activated to produce proteins for body functions.
- **Ghrelin**: The "I'm hungry" hormone that spikes before meals.
- **Glycation**: When sugars stick to proteins, causing damage and aging.
- **Gratitude Practice**: Regularly noting things you're thankful for to boost mood and health.
- **GrimAge**: A DNA test predicting disease and death risk more accurately than cholesterol.
- **Gut-Brain Axis**: The communication link between your gut and brain, affecting mood and health.
- **Heat-Shock Proteins (HSPs)**: Cell helpers activated by heat, like saunas, to repair proteins.
- **High-Intensity Interval Training (HIIT)**: Short bursts of intense exercise followed by rest.
- **Hormesis**: Small stresses, like fasting or cold, that make your body stronger.
- **HRV (Heart-Rate Variability)**: Variation between heartbeats; higher means better recovery.
- **Hyperbaric Oxygen Therapy**: Breathing pure oxygen in a pressurized chamber for healing.
- **Ikigai**: Japanese concept of your life's purpose, driving motivation.
- **Inflammaging**: Chronic low-grade inflammation linked to aging.
- **Inflammatory Markers**: Blood indicators like C-reactive protein showing inflammation levels.
- **Insulin Sensitivity**: How well your body uses insulin to control blood sugar.
- **Intermittent Fasting**: Cycling between eating and fasting to improve metabolism.
- **Interoception**: Sensing internal body signals, like hunger or heartbeat.
- **Ketone**: Backup fuel made by your liver when carbs are low.
- **Laughter Therapy**: Using humor to improve mood and physical health.
- **Leptin**: The "I'm full" hormone that signals satiety.
- **Lipid Profile**: Blood test measuring cholesterol and triglycerides for heart health.
- **Longevity Genes:** Genes like FOXO3 linked to longer lifespans.
- **Meditation**: Focusing the mind to achieve calm and clarity.
- **Metformin**: Diabetes drug studied for anti-aging benefits.
- **Mindfulness**: Being fully present without judging the moment.
- **Mitochondria**: Cell power plants producing energy for all functions.
- **mTOR**: A cell growth switch; balancing it is key for longevity.
- **NAD^+**: Molecule for energy and DNA repair, declining with age.
- **Neurogenesis:** Creating new brain cells, boosted by exercise and learning.
- **Neuroplasticity**: Brain's ability to rewire and form new connections.
- **Neurotransmitters**: Chemicals like serotonin carrying brain signals.
- **NMN (Nicotinamide Mononucleotide)**: Boosts NAD^+ for cellular energy.
- **Omega-3 Fatty Acids:** Essential fats in fish and flaxseeds for heart and brain health.
- **Oxidative Stress**: Imbalance between free radicals and antioxidants, harming cells.
- **Oxytocin**: Bonding hormone from hugs and trust, boosting well-being.
- **Partial Reprogramming**: Making cells younger without causing issues.

- **Placebo Effect**: Feeling better from believing in a treatment, even if it's fake.
- **Polyphenols**: Plant compounds acting as small stresses to slow aging.
- **Positive Affirmations**: Statements to challenge negative thoughts and build confidence.
- **PQQ**: Nutrient helping grow new mitochondria.
- **Prebiotics**: Fibers feeding beneficial gut bacteria.
- **Prefrontal Cortex**: Brain area for decision-making and self-control.
- **Probiotics**: Live bacteria supporting gut health and digestion.
- **Proteostasis**: Maintaining proper protein folding and function in cells.
- **Psychoneuroimmunology**: Study of how the mind affects the immune system.
- **Purpose in Life**: Having a sense of meaning driving your actions.
- **Quercetin**: Plant compound with antioxidant and anti-inflammatory effects.
- **Rapamycin**: Drug slowing cell growth, studied for longevity.
- **Red-Light Therapy**: Using 660 nm light to boost cell energy.
- **Resilience**: Bouncing back quickly from stress or challenges.
- **Resveratrol**: Red wine compound with potential anti-aging effects.
- **Sarcopenia**: Age-related muscle loss, preventable with exercise.
- **Senescent Cell**: "Zombie" cell that stops dividing and causes inflammation.
- **Senolytics**: Drugs or nutrients clearing senescent cells.
- **Sirtuins**: Longevity proteins activated by fasting or resveratrol.
- **Sleep Hygiene**: Habits like consistent bedtimes for better sleep.
- **Social Connection**: Relationships with others for emotional support.
- **Spermidine**: Compound promoting autophagy, the cell's cleaning process.
- **Stem Cell Therapy**: Using stem cells to repair damaged tissues.
- **Strength Training**: Exercises like weightlifting to build muscle.
- **Stress Management**: Techniques like meditation to reduce stress.
- **Synaptic Plasticity**: Changes in synapse strength for learning and memory.
- **Telomere**: Protective DNA caps that shorten with age.
- **Urolithin A**: Gut-made compound cleaning old cell parts.
- **Vagus Nerve**: Nerve linking brain and body, aiding relaxation.
- **Visualization**: Imagining desired outcomes to boost motivation.
- **VO_2 Max**: Maximum oxygen use during exercise, predicting longevity.
- **Wearable Technology**: Devices like smartwatches tracking health metrics.
- **Zone-2 Training**: Easy cardio burning fat and boosting mitochondria.

How to Use These Appendices

Here's how to make the most of these tools—like a quick-start guide for your longevity journey:

• **Appendix A**: Print or digitally fill out the Baseline Lab & Metric Sheet before Day-1. Repeat on schedule (Day-30, Day-90) to track your progress.

• **Appendix B**: Pick supplements from the Cheat Cards based on your goals and lab results—start with the Core Daily Stack, and always check with your doctor for peptides.

• **Appendix C**: Use the Action Worksheets as weekly checklists—one page per chapter. They're your "Do-Now" guide to make each chapter's lessons stick.
• **Appendix D**: Explore the Resource Directory to find trusted labs, gear, podcasts, and communities to support your journey.
• **Appendix E**: Scan the Glossary whenever you hit a big word—it's here to make learning easy and friction-free.
Final Reminder: Flip back to Appendix A now, log your Day-0 numbers, and set tomorrow's start time. Your "forever switch" activates the moment pen meets paper.

"Live as if you were to die tomorrow. Learn as if you were to live forever." – Mahatma Gandhi

PLEASE LEAVE A REVIEW

Now that you have everything you need to **work towards a longer, healthier life**, it's time to share your newfound knowledge and show other readers where they can find the same support.

By leaving your honest opinion of this book on Amazon or wherever you purchased it, you'll help others discover the guidance they need to elevate their voices and share their passion for **a healthy, meaningful, long life.**

Thank you for your help. The **quest for answers** lives on when we pass on what we've learned, and you're helping me to do just that.

If you purchased my book on Amazon, please leave your review.

REFERENCES

1. LICENSE-LINK REFERENCES

Creative Commons. (n.d.). ATTRIBUTION-SHAREALIKE 4.0 INTERNATIONAL (CC BY-SA 4.0) [LICENSE]. https://creativecommons.org/licenses/by/4.0/
Creative Commons. (n.d.). ATTRIBUTION-SHAREALIKE 2.0 INTERNATIONAL (CC BY-SA 2.0) [LICENSE]. https://creativecommons.org/licenses/by/2.0/

2. BOOKS & MONOGRAPHS

Sugrue, T. (1942). THERE IS A RIVER: THE STORY OF EDGAR CAYCE. Henry Holt.
Targ, R., & Puthoff, H. E. (1977). MIND-REACH: SCIENTISTS LOOK AT PSYCHIC ABILITY. Delacorte.

3. PEER-REVIEWED JOURNAL ARTICLES

Bishop, D. J., Botella, J., Genders, A. J., Lee, M. J-C., Saner, N. J., Kuang, J., Yan, X., & Granata, C. (2019). High-intensity exercise and mitochondrial adaptations. PHYSIOLOGY, 34(4), 243–255. https://doi.org/10.1152/physiol.00038.2018

Chen, S.-N., & Yang, C.-M. (2018). Macular telangiectasia: Pathophysiology and management. INTERNATIONAL JOURNAL OF MOLECULAR SCIENCES, 19(11), 3098. https://doi.org/10.3390/ijms19113098

Epel, E., Daubenmier, J., Moskowitz, J. T., Folkman, S., & Blackburn, E. (2009). Can meditation slow cellular aging? ANNALS OF THE NEW YORK ACADEMY OF SCIENCES, 1172, 34–53. https://doi.org/10.1111/j.1749-6632.2009.04355.x

Fischer, D., Lombardi, D. A., Marucci-Wellman, H., & Roenneberg, T. (2017). Chronotype and length of daylight independently predict sleep timing in humans. CURRENT BIOLOGY, 27(4), 572–579. https://doi.org/10.1016/j.cub.2016.12.015

Fitzgerald, K. N., Hodges, R., Hanes, D., Stack, E., Cheishvili, D., Szyf, M., … Bradley, R. (2020). Potential reversal of epigenetic age with diet and lifestyle. MEDRXIV. https://doi.org/10.1101/2020.07.07.20148098

Houghton, C. A., Fassett, R. G., & Coombes, J. S. (2016). Sulforaphane and cardiovascular health. PHYTOCHEMISTRY REVIEWS, 15(3), 281–295. https://doi.org/10.1007/s11101-015-9435-1

Kox, M., van Eijk, L. T., Zwaag, J., van den Wildenberg, J., Sweep, F., van der Hoeven, J., & Pickkers, P. (2014). Voluntary activation of the sympathetic nervous system and attenuation of the innate immune response in humans. PROCEEDINGS OF THE NATIONAL ACADEMY OF SCIENCES, 111(20), 7379–7384. https://doi.org/10.1073/pnas.1322174111

Morris, M. C., Tangney, C. C., Wang, Y., Sacks, F. M., Bennett, D. A., & Aggarwal, N. T. (2015). MIND diet associated with reduced incidence of Alzheimer's disease. ALZHEIMER'S & DEMENTIA, 11(9), 1007–1014. https://doi.org/10.1016/j.jalz.2014.11.009

Moseley, J. B., O'Malley, K., Petersen, N. J., Menke, T. J., Brody, B. A., Kuykendall, D. H., … Wray, N. P. (2002). A controlled trial of arthroscopic surgery for osteoarthritis of the knee. THE NEW ENGLAND JOURNAL OF MEDICINE, 347(2), 81–88. https://doi.org/10.1056/NEJMoa013259

Nordgren, T. M., Lyden, E., Anderson-Berry, A., & Hanson, C. (2017). The impact of vitamin D on immune function. NUTRITION IN CLINICAL PRACTICE, 32(6), 792–804. https://doi.org/10.1177/0884533617742820

Ravussin, Y., Xiao, C., Gavrilova, O., & Reitman, M. L. (2014). Effects of intermittent cold exposure on brown adipose tissue and metabolic parameters. JOURNAL OF CLINICAL INVESTIGATION, 124(2), 817–831. https://doi.org/10.1172/JCI71319

Takahashi, K., & Yamanaka, S. (2006). Induction of pluripotent stem cells from mouse embryonic and adult fibroblast cultures. CELL, 126(4), 663–676. https://doi.org/10.1016/j.cell.2006.07.024

Donovan, C. M., & Brooks, G. A. (1983). *Endurance training affects lactate clearance, not lactate production.* American Journal of Physiology: Endocrinology and Metabolism, 244(1), E83–E92. https://doi.org/10.1152/ajpendo.1983.244.1.E83

Long, J. (2014). Near-death experiences: Evidence for an afterlife? MISSOURI MEDICINE, 111(5), 372–377. https://www.ncbi.nlm.nih.gov/pmc/articles/PMC6172100/

Gershman, S. J., & Uchida, N. (2019). Believing in dopamine. *Nature Reviews Neuroscience, 20*(11), 703–714. https://doi.org/10.1038/s41583-019-0220-7

4. MAGAZINE, NEWSPAPER & POPULAR-MEDIA ARTICLES

BBC Future. (2016, May 10). Real-life superhuman strength: How people can lift cars. https://www.bbc.com/future/article/20160510-superhuman-strength-lifting-cars

Champion, C. (2024, November 18). Fast-track your health: Inside the fasting-mimicking diet. UCLA HEALTH. https://uclahealth.org/news/article/fast-track-health-inside-fasting-mimicking-diet

Godman, H. (2014, April 9). Regular exercise changes the brain to improve memory and thinking skills. HARVARD HEALTH PUBLISHING. https://www.health.harvard.edu/blog/regular-exercise-changes-brain-improve-memory-thinking-skills-201404097110

Hagan, E. (2022, January 21). The science of novelty. PSYCHOLOGY TODAY. https://www.psychologytoday.com/us/blog/healing-stress-the-inside-out/202201/the-science-novelty

Lowery, G. (2010, December 6). Study looks at the brain's ability to see the future. CORNELL CHRONICLE. https://news.cornell.edu/stories/2010/12/study-looks-brains-ability-see-future

Svoboda, E. (2006, July 1). Pay it forward. PSYCHOLOGY TODAY. https://www.psychologytoday.com/us/articles/200607/pay-it-forward

5. SCIENTIFIC & MEDICAL NEWS RELEASES / NIH & HARVARD HEALTH SITES

Harvard Health Publishing. (2024, March 28). How to boost your immune system. https://www.health.harvard.edu/staying-healthy/how-to-boost-your-immune-system

National Institutes of Health. (2016, August 16). Epigenetics, diet and lifestyle. https://www.nih.gov/news-events/nih-research-matters/epigenetics-diet-lifestyle

National Institutes of Health. (2018, July 17). Eliminating senescent cells extends healthy life in mice. https://www.nih.gov/news-events/nih-research-matters/eliminating-senescent-cells-extends-healthy-life-mice

Lewis, K. (2021, April 27). GUIDED VISUALIZATION: DEALING WITH STRESS. National Institute of Mental Health. https://www.nimh.nih.gov/news/media/2021/guided-visualization-dealing-with-stress

Tafur, J., Van Wijk, E. P. A., Van Wijk, R., & Mills, P. J. (2010). *Biophoton detection and low-intensity light therapy: A potential clinical partnership.* **Photomedicine and Laser Surgery, 28**(1), 23–30. https://doi.org/10.1089/pho.2008.2373

6. ORGANIZATIONAL & DATABASE-LEVEL URLs

American Psychological Association. (n.d.). STRESS & SOCIAL SUPPORT. https://www.apa.org/topics/stress/social-support

HeartMath Institute. (n.d.). RESEARCH LIBRARY. https://www.heartmath.org/research/research-library/

JAMA Network. (n.d.). https://jamanetwork.com/

National Library of Medicine. (n.d.). PUBMED. https://pubmed.ncbi.nlm.nih.gov/

Nature Portfolio. (n.d.). EPIGENETIC CLOCKS MEASURE BIOLOGICAL AGE. https://www.nature.com/articles/s43587-021-00094-7

New England Journal of Medicine. (n.d.). https://www.nejm.org/

Wiley Online Library. (n.d.). https://onlinelibrary.wiley.com/journal/14749726

7. WEBSITES, BLOG POSTS & MISCELLANEOUS ONLINE SOURCES

Gaines, J. (2020, November 17). Ikigai: A Japanese concept for finding purpose in life. POSITIVEPSYCHOLOGY.COM. https://positivepsychology.com/ikigai/

Moore, C. (2025). Managing a values-actions gap. THRIVE STREET ADVISORS BLOG. https://www.thrivestreetadvisors.com/blog/managing-a-values-actions-gap

Richards, M. (2023, May 3). Three ways getting outside helps improve your health. UC DAVIS HEALTH BLOG.
https://health.ucdavis.edu/blog/cultivating-health/3-ways-getting-outside-into-nature-helps-improve-your-health/2023/05

Stillman, J. (2021, August 3). Elon Musk's favorite brain-training trick. INC.
https://www.inc.com/jessica-stillman/elon-musk-brain-training.html

U.S. Air Force National Museum. (n.d.). BATAAN DEATH MARCH.
https://www.nationalmuseum.af.mil/Visit/Museum-Exhibits/Fact-Sheets/Display/Article/196797/bataan-death-march/

Woo, J. (n.d.). MALALA'S QUIET ACTIVISM FOR EDUCATION AND PEACE. Vision of Humanity. Retrieved May 7, 2025, from https://www.visionofhumanity.org/malalas-quiet-activism-for-education-and-peace/

Carpenter, J. (2016, August 30). QUEEN OF PEACE, MEDJUGORJE, BOSNIA-HERZEGOVINA (1981–PRESENT). Divine Mysteries. https://www.divinemysteries.info/queen-of-peace-medjugorje-bosnia-herzegovina-1981-present/

BrainyQuote. (n.d.). *Assorted quotations on motivation, health, and success.* Retrieved May 7, 2025, from https://www.brainyquote.com/

ABOUT THE AUTHOR

Tad Sisler is an American Composer, Author and Producer of feature films and music. More than a thousand of his original works are available through *iTunes, Amazon* and virtually every other major marketplace. Through the years, **Tad** created and released independent feature films and documentaries, television shows, developed a music store and vast collection of music for film and television usages, in addition to published screenplays and books.

Tad is a voting member of *The Academy of Recording Arts & Sciences.* **Tad** invented a wireless karaoke all-in-one microphone that became a best-seller on *Amazon.* A child prodigy, Tad was playing advanced piano pieces at the age of 8, and rating superior in Classical piano competitions at 12. Tad won his first scholarship for singing at 12, attending the Idyllwild School of Music and the Arts, then affiliated with the University of Southern California.

FEATURE FILMS

Tad produced, edited, and released "**The Ghosts of Brewer Town**", a mystery feature film, currently available on *YouTube.*

TELEVISION PROJECTS

Tad launched the **Journey To An Extraordinary Life-Legends Among Us** documentary series, which chronicles the lives and careers of legendary artists, actors, sports figures and heroes of medicine, in a feature-film format.

BOOKS

Books, Audio Books and Podcasts released by **Tad** include **"Reflections in the Key of Life-The Steve Madaio Story",** chronicling the life and times of America's most prolific trumpeter.

This book garnered a **Readers' Favorite Book Award** for Tad.

"Mafia Baby" is a shocking true story of a woman raped by a Mafioso, who then raised his child alone. Tad's autobiography, **"It's a Long Climb to The Middle"** ***is*** available currently on *Amazon* and *Barnes & Noble*. Screenplays in development by Tad Sisler include **"The Incredible Spark of Franklin Benjamin"**, and **"Please Don't Forget"**. **Tad's** latest **Music Mastery** collection of books is designed to educate and inspire musicians to become masters. His **Health and Longevity Mastery** series of books is crafted to educate on longevity, age reversal, and general wellness.

MUSIC

Tad's production music catalog tripled in size with the addition of thousands of excellent production music tracks, as well as hundreds of sound-alike tracks for the DJ/Karaoke industry, now distributed on **iTunes, Amazon Marketplace, CD Baby, Spotify, Rdio, Xbox Music** and dozens of other outlets Worldwide.

Tad produced and released "The Barcelona Sessions" to 1000 radio stations Worldwide, with never-before-heard original performances by Miles Davis' bassist, Bill Evan's drummer, Frank Sinatra's saxophonist, Maynard Ferguson's guitarist, and Andrae Crouch' flutist/saxophonist, produced by Tad Sisler in his recording studio.

Tad Sisler composed the full score to "**The Encore Of Tony Duran**", an indie feature film starring **Elliott Gould, William Katt, Nicki Ziering and Cody Kasch**, along with his co- composer Andrew Fraga, Jr. After having the distinction of being the first film to sell-out at the prestigious *Palm Springs International Film Festival*, the film won the **Jury Award** for **Best Feature Film** at the *Las Vegas Film Festival* and the *Santa Fe Film Festival*, as well as the **Indie Spirit Award** at the *Fort Lauderdale Film Festival* and the **Audience Favorite Award** at *Tallgrass Film Festival*, in conjunction with a **Lifetime Achievement Award** for **Elliott Gould.** The film is available on *Amazon Prime.*

Tad completed the music and audio editing for the TV Series "**American M.C.**". The first 7 episodes are complete and in the process of distribution through **iTunes**. Tad scored the Main Title theme to **American M.C.** as well as underscore and providing Music Supervision and source music.

PRODUCTION

Tad Sisler has been a valuable member of the team of specialists and project developers for **Yamaha Corporation of America**, delivering hundreds of intricate projects to exact **Yamaha** specifications over a 10-year period.

Tad received accolades in 2011 after being given the honor and challenge of doing the "official" remake of the iconic "**Andy Griffith Theme**" for the estate of the composer **Earle Hagen** as a perfect sound-alike, along with his composing associate Andrew Fraga, Jr.

Following a stint composing for a series entitled **"Famous Families"** on **Foxstar** and working as assistant to composer Jeff Edwards on the television series "**Silk Stalkings**" and "**Renegade**" in the late 1990's, Tad Sisler and founded & developed a production music catalog, containing thousands of high-quality music tracks available for sync licenses in film, television and advertising in more than 150 genres.

In addition to handling Music Supervision on "**The Encore Of Tony Duran**", and on **"American M.C."**, "**The Ghosts of Brewer Town"**, **"Tis' The Season"**, the **"Journey To an Extraordinary Life"** series, **Tad** placed his original music on **NBC, ABC/Disney, Warner Brothers Television, TNT**, US National Infomercial campaigns through **Guthy/Renker** and **Script To Screen**, as well as custom composing for the TV and Advertising industry.

Tad released contains hundreds of top-quality soundalike tracks produced by **Tad** and his associates, for DJ and Karaoke usages, currently on *ITunes, Amazon Marketplace, Spotify, Rdio, Xbox Music,* and many other outlets.

LIVE PRODUCTION

In the 1980's and 1990's, **Tad** and his team produced a series of live headliner events at multiple venues from the ground up, including sold-out performances by **Kenny Rogers, Earth, Wind & Fire, Los Lobos, Glen Campbell, The Righteous Brothers, Lou Rawls, Tito Puente,** the **Power Jam** featuring **Timmy T, Tara Kemp, Candyman, Soul To Soul** and more.

HISTORY

As a very young man, Tad Sisler worked as a performer for **Frank Sinatra**, studied music in choreography under world-famous Broadway Dancer/Choreographer **Jacque D'Amboise**, received superior ratings in classical piano performance in tough **Joanna Hodges** international competitions, and received private acting lessons from **Richard Burton**, a friend of his family.

Tad attended the prestigious **Idyllwild School of Music and the Arts** on vocal music scholarships during the period when it was affiliated with the **University of Southern California**. In High School, Tad was one of 100 statewide vocalists elected to the prestigious **All-State Choir** in Missouri.

During his storied career, Tad has also had the honor of performing with and working among such greats as **Gladys Knight, Rita Coolidge, B.B. King, Marilyn McCoo, Johnny Mathis, Kenny Rogers, Tito Puente, Sonny and Mary Bono, Gene Barry, Terry Cole-Whittaker, Shecky Greene, Peter Marshall, Mary Hart, Blackwell, Herb Jeffries, Trini Lopez, Glen Campbell, Jennifer Hudson** and other legends.

Tad Sisler's extensive experience, state of the art facility and history of delivering quality feature films and music on time and on budget, as well as the ability to draw from an extensive catalog of production music, allows his experienced team to offer complete services in custom film and television production as well as in music composition and production efficiently.

Tad is proud and humbled to be a voting member of the **Academy of Recording Arts & Sciences**, which allows him to have a voice to vote for great artists worthy of winning a **Grammy Award**. Many of Tad's works have been placed into Grammy consideration.

In 2023, Tad won a prestigious **Telly Award** for creative excellence in his *Journey to an Extraordinary Life* film series.

Modern Renaissance Publishing is at the forefront of a new intellectual awakening, dedicated to fostering a renaissance of ideas that resonate in today's world. Our mission is to bring cutting-edge concepts and timeless wisdom to the public through a diverse array of publishing formats, including books, eBooks, and audiobooks.

We are proud to launch our **Music Mastery** series, offering comprehensive guides and insights for musicians of all levels. Our **Health and Longevity Mastery** series highlights the latest discoveries and insights into extending the human healthspan and lifespan. In addition to our literary endeavors, we also publish original music, enriching the cultural landscape with creative expressions. Whether you're seeking to expand your knowledge, enhance your skills, or simply be inspired,

Modern Renaissance Publishing provides the resources and content to empower your journey. Join us as we bridge the rich heritage of the past with the innovative spirit of the present to shape a brighter, more enlightened future.

Publisher: MODERN RENAISSANCE PUBLISHING
IN USA +1 (818) 845-6700
modernrenaissancepublishing.com
Email: modernrenaissancepublishing@gmail.com
ISBN#978-1-966258-21-6

MODERN RENAISSANCE
PUBLISHING

Made in the USA
Monee, IL
04 November 2026

67387604R00085